C000142046

The

BIG

And Easy Guide

To Take a Bright Idea from
Drawing Board to Successful Revenue

Published by Jembro
England

Acknowledgement

With special thanks to

Stefan Viklund

*for helping make this
book possible*

CONTENTS

Foreword
> Dr John Beacham CBE; DSc; FRSC

Introduction
> Ideas and the Black Hole theory

Chapters

Foreword

In my time in international industry, I have seen many good ideas come to fruition, some of which were global successes; but many have disappeared off the face of the earth.

I've often asked myself the question "Where did they go to?" and more importantly "Why?" I believe the main cause from my observations is that many in industry, particularly in the SME sector are not equipped with the required skill sets to successfully exploit such creativity. Indeed supporting my observations are several government reports that highlight this issue.

The British are famed for their inventiveness, yet turning this into sustainable, successful income does not happen anywhere near as much as it should.

So how do you get a notional, creative idea through the many stages of its development cycle to realise commercial success? It is not an easy process by any means.

If you have an idea, then Rob's book is an essential read, explaining in an uncomplicated way the whole process, from that first "light bulb" moment through each stage of the development cycle with one aim - achieving commercial success.

I wish you every success in your endeavours.

Dr John Beacham
CBE; DSc; FRSC

Introduction

If there is one thing each of us is good at, it's coming up with ideas, whether it's through an inspired moment or through intense research. If there's one thing we're bad at, it's doing something with them and even if we do, we can spend a considerable sum on costly mistakes, often into the thousands of pounds. This book will show you how to avoid those costly mistakes and demonstrates ways in which you can capitalise on your idea.

There are many varied statistics about how often we think about sex (one report quotes every seven minutes!) and I don't think our creation of ideas, many of which are fleeting, is any different. We're not talking ideas of a sexual nature here, but ideas that solve a problem, ideas which create new opportunities and ideas which are sometimes so simple, that they are dismissed because of the simplicity of the notion.

The human brain is indeed fertile ground in which to sow the seeds of ideas; ideas about which we are all inspired to think of – that creative moment of world changing genius. It could be simply prompted by watching a news item on television, inspired by spotting something whilst walking along the street, observing how things are done in the work place or lab, maybe something someone said about a specific problem or that bolt out of the blue eureka moment!

Whatever the inspiration behind that "Idea Moment", one thing is for certain, we have them in abundance and quite frankly, on nearly all occasions, we do absolutely nothing with them!

At the turn of the Millennium an amazing statistic from a survey carried out by the Japanese Department of Trade in its review of the 20th Century stated:

"54% of the most important and influential inventions in the previous 100 years were British."

It is staggering to believe that Britain invented more than half of those inventions and that the rest of the world together, including the hugely powerful and influential nations of the East and the West, accounted for a collective, and somewhat paltry 46%.

Those statistics say a huge amount about the awesome creativity of the British.

Britain has been called a nation of inventors, but one thing is for sure, we may have invented, but we didn't manufacture or commercialise anything near that percentage, yet the rest of the world capitalised on that genius.

This lack of capitalising on our inventiveness is still a trait that is as relevant today as it was in the 20th Century. So where do we go wrong? Put simply, it's the Innovation Black Hole, a woeful void into which most ideas, creativity and inspiration are drawn, never again to see the light of day.

This book paves the way in avoiding that Black Hole and demonstrates how to look at an idea drawn up on a piece of scrap paper with the knowledge that it could possibly be turned into the next "must have" product, selling around the world.

Each stage of the idea development and management process is explained in an uncomplicated, easy to understand style, allowing you to avoid many of the pitfalls along the way, so giving your idea the best opportunity to progress through to successful commercial exploitation.

One thing is for sure, many of your ideas won't make it, but using the processes in this book, you will know which ones won't, allowing you to concentrate on the ones that will.

The book is intended as a useful tool for individuals, whether in small businesses, large corporate organisations, universities or other institutions where an individual or individuals come up with that novel idea. The process is also designed to allow banks, financial institutions and business intermediaries to make considered judgements when investing in or helping their clients' to develop and commercialise their idea.

Each chapter is designed to overcome any problems that may arise in the idea development and management process, pre-empting easy yet sometimes very costly mistakes, thus minimising the risk and maximising the opportunity.

Innovative ideas in processes and services are not excluded, but there is a focus on those innovative ideas of a practical, physical nature, where development of such an idea would typically lead to protection of the intellectual property by way of a patent and maybe a registered design.

Whatever you do, when you have an idea, even for a fleeting moment, don't discount it! Capture it and file it away, then return to it sometime later after you've had further time to think about it.

Over the years, I have seen my fair share of barking mad ideas through to those that are brilliant, yet I have always treated them equally. Do that with your own ideas; they may appear barking mad or glaringly obvious, but equally, they could just turn out to be that one moment of brilliance that could set you on the road to a comfortable income.

I trust that this book will help you discover if that's the case.

Rob Lucas

Chapter One

Setting Out

The golden rule

There's one golden rule in having an idea and working to develop and commercialise it, so let's start off as we mean to go on:

Be invitational, not confrontational!

It may seem an odd thing to say from the outset, but your idea is yours; it's your precious idea and it is natural to think woe betide anyone who comes between you and your idea.

Throughout the development of your idea, you will call upon relevant experts who with best endeavours will be trying to help you succeed. Seek their counsel wisely and invite their criticism, which may appear good or bad, so long as it's constructive. Adopting this approach is one of the greatest virtues in idea development and will assist you in realising your ambitions for the idea.

The process

Throughout the course of this book, processes will evolve that will help you assess at each stage where your idea is headed.

It is important to realise that whilst the chapters in this book follow on from one to another, many of the processes explained may run concurrently. This is very much dependent on your idea and its natural path of development, so be prepared to "spin some plates" at the same time!

Let's be clear that the processes are intended to make sure that your weak ideas fail and that your good ones have the best chance of success. Harsh as that may sound, it is better to direct your valuable resource towards those opportunities that have a good chance of coming to fruition than pursuing something down a blind alley, incurring wasted time and cost. Yes, there will be a cost to developing your idea, but following the processes outlined in this book will help you to spend wisely and not end up on a financial "fool's errand".

The processes apply equally to ideas emanating from any individual, either on their own, in a Small to Medium Enterprise (companies with less than 250 employees), or in larger corporate organisations and institutions - universities included. Whilst each will have their own nuances and funding levels depending on their style of operation, the process and results obtained hold firm.

Your idea could be a product, a process or a service. It could be purely innovative, improving that which went before with no Intellectual Property (IP) attached, or it could be that ethereal moment of brilliance, full of IP that could be deemed to be innovative.

Innovation and Intellectual Property – The difference

Innovation and Intellectual Property have often been used in the same breath yet sometimes never the two shall meet. Innovation is a much overused word. Since the government launched their Innovation report in 2003 signalling their intent to make the UK a knowledge-based economy, innovation has been on the tip of everyone's tongue.

Innovation
Innovation can be described as:

- *Noun*
  A creation (a new device or process or service) resulting from study and experimentation

  The creation of something in the mind

  The act of starting something for the first time; introducing something new

Innovation applies to products, processes and services. It may be a way of doing something in a different fashion that is new and improves an already existing method. An example is rearranging the way machines on a factory floor are laid out and function together, along with changing the operational methods of the shop-floor staff so improving the efficiency of output. This would be considered an innovative approach, but it may not have any IP attached to it.

Innovation is change for the better that affects the way a product works, the method in which a process is implemented or the way in which a service is improved, of which some may or may not have generated IP.

Intellectual Property
IP can best be described as:

- *Noun*
  A product of the intellect that has commercial value, including copyrighted property such as literary or

17

artistic works and ideational property such as patents and registered designs

IP is generated by the origination of a thought which is both inventive and novel – in a nutshell, ideas! It's an idea that has originated from within you, inspired by events, observations or practical experience. If yours is an idea that isn't out in the big, wide world, it will be for one of three reasons:

- It's been tried before, failed miserably for a variety of reasons and then disappeared off the face of the earth

- Someone has thought of it in a fleeting moment of brilliance, but has never done anything with it

- It has never been thought of before

If your idea is the latter, then you are possibly onto something; your own IP that has resulted from your original creative thought. As soon as this is realised, you have property of your own intellect; an asset. It is of course very early in the process to see whether that asset has tangible worth or is original and if your idea progresses through each stage in this book, then your idea could just possibly be worth a considerable amount.

But before we get carried away and transported off into the universe of pound and dollar signs, you have to start at the very beginning, a stage which will allow very basic and fundamental checks to be undertaken at minimal or no cost.

Capturing your idea

So you've had what you believe to be a world beating, "eureka moment" thought. One very simple rule of thumb is:

If it stays in the head, it stays dead!

Once you've had that idea, of course mull it over for a bit, but at the very earliest opportunity, get it down on paper then file it away.

I have personal experience of two ideas that I thought of and at the time, neither recorded nor developed them. Lo and behold, they eventually both appeared in the marketplace; one ten years after I had the first thought and one some twenty five years later. I still reflect back to those "what if I'd......" moments. In these instances, hindsight is a wonderful gift; we've all been blessed with it.

The most important discipline at this very early stage is do not tell anyone, even your nearest and dearest, nor close friends or work colleagues. Although you've jotted it down on paper, you will still be rounding your thoughts and considering "what ifs?" If others know, their input at this stage will inadvertently divert your thinking from your prime objective; "could this work and have I thought of everything?" Of course you will not think of everything, but revisit your piece of paper and make additions as and where appropriate. This process may take a couple of days, a month or even longer before you get to that point of being reasonably satisfied that the idea has taken shape.

There will come a point where you will have looked at the practicality of your idea, and asked yourself:

• Could it possibly work?

- Does it address the issue or solve the problem that it should?

- Would I buy one?

If you've asked those questions and are moderately satisfied that your idea may answer those questions in a positive fashion, then now is the time to talk to someone else and see what they think in terms of those three questions that you've asked yourself. Rest-assured, whilst they may offer some constructive criticism, they will often over-enthuse about your idea. Why? Because they want it to work for you and all will convince you that they would buy one if your idea gets into the market place. Take this enthusiasm on board with gratitude, yet taint it with caution.

It is very important to note here that if you confide in someone else about your idea (unless it is a professional body that is legally bound by confidentiality), you are technically putting your idea into the public domain and in doing so, you could invalidate any patent application that you file. You should get a Confidentiality Agreement [(sometimes called a Non-Disclosure Agreement (NDA) or Confidential Disclosure Agreement (CDA)] signed by any person in whom you confide. Confidentiality Agreements are explained in Chapter Two; "Confidentiality – Keeping your idea out of the public domain".

Of equal importance at this very early stage is the question "is this going to be my idea alone and only I own it?" The reason to consider this is fairly straight forward; others to whom you show the idea in confidence will have input undoubtedly. They will share their thoughts with enthusiasm as to how to change or improve your idea, some of which may be fundamental to its possible future success. You have to decide which way you will jump on this one. If you intend all along that this is your idea alone and you own it, make that clear from the outset that whilst you appreciate their input, it will still be your idea alone. Conversely, you may decide that you need that level of input from others to have any chance of it really working and choose to include others and share the ownership of the idea. Only you can make that decision. If you decide that you alone own the idea, you can always change your mind at a later stage and include others if it suits the development of the idea. It is far more difficult to share the ownership from the outset and then change your mind further down the line. This is explained in detail in Chapter Six; "Who owns the idea? Tip-toeing through the minefield"

Recording Your Idea

Whatever you do, don't get grandiose dreams that your idea is the next multi-million pound opportunity. In reality these are very few and far between and one of the biggest and most costly mistakes anyone can make at this stage is rushing to file a patent application. It is natural to think that you have to protect it before the rest of the world copies it, but in reality, the chance that someone elsewhere in the UK or overseas has come up with exactly the same idea at the same time and then filed a patent application just ahead of you is remote. Your idea is at that very early stage and IP protection by way of a filed patent application or registered design should not, almost without exception, be considered at this point. Of course, if your idea progresses, some form of IP protection will probably be the key to its future commercial success, but not yet, as your idea will change throughout its development. Details of the different types of IP protection are explained in Chapter Five; "Protecting your idea".

A few years ago, I came across one such instance of filing a patent for an idea far too early. A small company came up with a novel idea for a drinks-heating apparatus. Their first design used a heating coil in the apparatus and to protect this idea, a patent application was filed, but such was their ambition, they did not just file for a patent application for the UK, they were looking to file worldwide.

Whilst pursuing the patent applications, over a two year period, the main element in the design changed from a heating coil to a webbing-based heating element. The heating coil was the main claim in their patents but was no longer viable on practical grounds, being replaced by the new heating element. The company had spent £12,000 on patent applications, trying to cover every corner of the globe, but now their patents were no longer relevant to the new design.

Their patent applications had to be abandoned as their claims did not contain the key element of the apparatus, the webbing-based heating element. They had to start all over again and file for a new patent application, having wasted the £12,000 filing previous applications far too early that included a heating element no longer relevant to the final product.

There is a very simple method to record the date on which you shaped your idea; not the date you first thought of it, but the date on which you feel reasonably comfortable that the idea has come together. Whilst this method of recording your idea carries little weight in a legal standing compared to a well written patent application or other forms of protection, it is a way to get a legally recorded date stamp against your IP.

When you have the definitive drawing(s) and notes finalised as best you can, sign and date them, put them in an envelope addressed to yourself, seal it and post it. When you receive it back complete with date stamp, DO NOT

OPEN IT! Sign or initial across the seal on the back and cover with transparent sticky tape. You now need to store this in a safe place either at home or preferably in the safe at a bank, solicitors or accountants (often they charge a small fee to do so). It may never be called upon as some form of proof, but at least you have a date on which you "filed" your idea.

From 2D to 3D – Making a working model

The following assumes that your idea can be or needs to be made as a physical, functional entity, yet whilst as a rule of thumb, most will fall into this category, not all do. Computer algorithms for example do not necessarily require a working model to be exploited by a commercial concern, so long as the theoretical principle holds fast and is proven, a commercial concern could well licence it for exploitation.

The thought processes so far have revolved around the idea of solving a problem, improving an existing product or creating something new in a totally different and novel way. The drawings and notes that you have made will seem to prove that the concept works and in some cases, depending on your idea, you will not need to do more than this at this stage.

In most cases, ideas do need to be practically proven and this is where the trusty "working model" comes into its own. Note that the working model is not the finished item, nor a prototype; it is the basic "proof of principle" demonstrating that the idea in practical terms can work. Important things to note:

• It doesn't have to be aesthetic and "look good"

- It does not have to be made of the materials in which the final product would be manufactured

- Size does not matter within reason – so long as the model works, it's proved its point. If the market (we'll go into market analysis in later chapters) dictates that it's got to be the size of a mouse and yours is the size of an elephant, you might have a problem convincing people!

- Don't over-engineer it with bells and whistles – you're trying to prove the key elements of your idea at this stage, that's all

One thing about completing a working model is that it is immensely satisfying! You've been able to prove that the original idea you had now works in its most basic form. What you will realise as soon as you have made it is that you can already improve it in terms of its functionality, size and materials. Only improve it if:

- It's not costing you much to do so

- It becomes apparent that an aspect or aspects of the working model such as a change of material or a way in which something operates can dramatically improve the functionality

- Eureka II occurs! – you've suddenly thought of something that if added, can dramatically improve the idea

The latter two of the three are the first examples in this book as to why you should never rush into filing a patent application. As can be seen, your working model is already evolving from the original notional idea that you had and one thing is for sure, it will continue to do so.

A reminder once again about confidentiality; there will be a temptation to show your model probably to more people than you originally shared the idea with under confidentiality. It's natural to want to do so as approval only endorses your belief in what you're doing; there is a triumphal sense in proving to yourself and others that it works. Put simply, DON'T DO IT!

The above assumes that you can in fact make a working model that functions as intended. This is not always the case as the materials or the physical complexity of the idea mean that it is not a straight forward idea from which you can make a working model. If from your drawings and notes, there are no obvious reasons why if a working model was built, it would not work, but to do so for you would be extremely difficult, there may be a good reason for seeking limited expert help at this very early stage in the overall process if this difficulty is a stumbling block. There will be a cost to this and may only be a few hundred pounds, but it could be a sound investment if it gets the working model functioning as desired. Emphasis here once again is on confidentiality. If you go down the route of seeking expert help, do not do so without a confidentiality agreement in place. See Chapter Two; "Confidentiality – Keeping your idea out of the public domain".

Sourcing this expertise can be done through various channels. Web searching is one obvious example, as well as word of mouth. There are various business support organisations that can signpost you towards the relevant expertise. Another invaluable source of relevant knowledge is universities. Searching a university's website can give you sufficient information as to whether they possibly have the relevant expertise. When on-line, look at the university's faculties that may have the possible relevant experience. Within the faculty will be schools which sub-divide into relevant specialisms, together with contact numbers. If they cannot assist, they will often point you in the direction of another faculty or university that can do so.

Oh No! Failure's on the cards

If your working model,

- Fails to be built by you because of complexities

- Is a high-risk project and costs a relative fortune to employ external expertise to make it work in its most basic form

- Is deemed not workable by experts in their field

Then now is the time to probably forget it!

This is the first stage of failure and it will be extremely disappointing, but you have kept your own time spent on the idea to a minimum and more importantly any costs incurred have been negligible. Your heart more often than not will say "continue because I know the world wants it!" I have seen many expensive mistakes made based on this misguided passion by following the heart and not listening to the head, but let the head have the casting vote and agree with its decision to bury the idea and move on to the next one.

On the upside, one good thing about having kept your idea confidential to a very few people is that you don't get that often-asked question "what happened to that idea of yours?" and "How are you getting on with it?" time and time again!

Oh Yes! You could be on your way

Your working model appears in fine order as best it can at this juncture – it proves the fundamentals of your idea in a practical and pragmatic fashion. Something which does the job and who knows at this stage, maybe more!

Very few know about it and those that do are next of kin, joint-inventors or others all under a code of confidentiality. This chapter started with "invitational, not confrontational" and this will become even more of an important factor beyond this point.

Be prepared to see your idea possibly change beyond all recognition as it jumps through the various hoops. So long as you or your co-inventors make sure that the IP is owned by you or all of you throughout the whole process, then accept any changes suggested by the relevant experts with grace. As they will charge you for their input into your idea, they will not normally lay any claim to your IP, but always check with them and make sure that the changes or additions that they make to your idea belong to you as your IP.

During the development of your idea, be prepared to give your input by way of your own suggestions although sometimes they may not agree, but I have always worked on (and experience proves it) that they are the specialists; they're doing it in the best interests of you and your idea as well as their own reputation.

Chapter Two

Confidentiality – Keeping your idea out of the public domain

The first important document that needs to put in place at the outset is the confidentiality agreement. This chapter is not intended to go into the fine legal detail of a confidentiality agreement. It is intended to give you a rounded understanding of such a document, what to look out for and how to avoid simple, yet possibly costly mistakes. A confidentiality agreement is often known as:

NDA - Non Disclosure Agreement
CDA - Confidential Disclosure Agreement

So why is it needed? The confidentiality agreement is an agreement that is entered into by two or more parties to ensure that disclosures, discussions and collaboration between those parties remain confidential. It is normally related to a specific subject; in this case, it would be your idea.

The parties

Who are the Parties? Obviously one of the parties will be you and your colleagues or your company. The other parties with whom you enter into a confidentiality agreement will depend on how far your idea develops, the nature of the idea and to whom it is shown. These could include industrial technical experts, universities, finance houses, colleagues, potential licensees and other parties to whom you may have to reveal your idea.

A correctly drafted confidentiality agreement is a legally enforceable document in a court of law whereby if one party breaches the terms of that confidentiality, the other may take legal action against such breaches. It is important to note here that it may be very difficult to prove that any breach has taken place unless it is evidentially apparent. Even if this is so, you have to consider if the breach has had a material effect on the development, commercialisation or reputation of the idea and even taking that into consideration, what the cost would be to pursue a legal action against the other party. A breach will often be down to human error, by letting something slip and often of a minor nature, although serious intentional breaches do occur from time to time.

I personally have yet to come across any legal action by one party against another for breach of confidentiality, although that is not to say that it doesn't happen. It is best to accept that by entering into a confidentiality agreement with another party, you are entering into the "spirit" of the agreement. In other words both parties are showing the spirit of intent and willingness to disclose, discuss and collaborate on a confidential basis. Once in place, it still does not stop one party breaching that agreement, but it is a risk mitigation exercise.

In respect of public domain disclosure, it is considered that the disclosure of your idea under a correctly drafted confidentiality agreement does not bring the idea into the public domain and still allows for you to file a patent application. This is another good reason for putting a confidentiality agreement in place with another party to whom you will disclose your idea, particularly if filing a patent application is under consideration (see Chapter Five; "Protecting your idea").

What is covered by a confidentiality agreement?

When entering into a confidentiality agreement, it is important to decide what is to be covered in respect of your idea. If the description of your idea in the agreement is very narrow, you may find that whilst working with the other party, developments to the idea may stray outside that which is covered in the agreement. Conversely if it is too broad and covers everything that you have ever thought of, then you could find that you are locked into confidentiality with the other party on just about everything including the kitchen sink! If you have not filed a patent application or registered design at this stage, then your description should include:

a) A title
b) What purpose the idea serves
c) Any particular features specific in functionality

An example of such:

a) A "Shoe Sole Coating"
b) Reducing wear on soles of shoes
c) Using a chemical application and heat treatment

This can be written into one or two sentences and would be broad enough, making reference to:

"Shoe sole coatings covering any sort of reduction of sole wear; using a broad range of chemicals and various types of heat treatments".

If a patent application or registered design has been filed for your idea, then the patent application number or the design registration number may be sufficient to enter into the confidentiality agreement as it could be considered that these applications, if well written, would include all the relevant detail about your idea.

Some companies, particularly large corporates or institutions will insist on having a patent application filed by you before they will even enter into a confidentiality agreement and require that the patent number of your idea is cited in their confidentiality agreement. Such companies are not usually interested in registered designs.

Whose agreement do you use?

This question generally falls into two distinct camps; yours or theirs. Larger companies, particularly corporates and institutions will insist that their agreements are used and there are two good reasons why this will be the case:

- Their agreements have been developed over many years and have stood the test of time and whilst they are designed to protect them, they are in general there to protect you also

- They have a legal affairs department that is issuing hundreds if not thousands of these agreements each year. It is neither practical in time nor money for them to consider other parties confidentiality agreements

You will have to accept that you will enter into their confidentiality agreement and not yours. If you don't agree to that, they will not be reviewing your idea!

Apart from the example explained above, if you have your own confidentiality agreement and only if it is well drafted, offer it as the agreement to be used; it shows that you are serious about your idea. It comes across that you are taking professional advice and that you are being diligent about each aspect of your idea's development. It may be refused by even the smallest of companies and specialists, but if that is the case, and they insist on using theirs, you can refuse to enter into their agreement and then seek companies and specialists elsewhere to assist you in your idea. Alternatively if you do use theirs, make sure that you check it through properly, that it is professionally drawn up and that your vested interests in terms of confidentiality are protected.

Whilst the intent in every confidentiality agreement is generally the same, each one will be different to some extent in the way it is worded and what it covers. What is important to remember is that a confidentiality agreement is generally a two-way document, although there are one-way agreements which protect the interests of one party only. In a two-way confidentiality agreement, whilst you may be only interested in protecting your interests and there is nothing wrong in that, it has to be considered that the other party, by entering into the agreement is protecting their own interests as well. If the party with whom you're entering into the agreement is amongst other things a professional, or a technical specialist, they will on many occasions give you an insight into their particular ways of working along with their specialist knowledge which is necessary to develop your idea.

This may well be their proprietary asset and they will want to make sure that you are bound by confidentiality, so legally preventing you from passing on how they work to other parties outside the agreement. This is standard in a confidentiality agreement, so do be careful what you say to any other parties outside of the agreement otherwise you could end up in breach.

Patent attorneys have a professional code of confidentiality, so there is no reason to insist upon a confidentiality agreement being in place before disclosing your idea.

There will be times when the other party to the agreement will require external help in evaluating or developing your idea; this is quite normal. Ensure that the agreement that you and the other party enter into covers these external parties and if it doesn't, insist that the other party enters into a confidentiality agreement with any party who is external to the original agreement and that you require proof that they have done so.

Whose name goes on the confidentiality agreement?

This is extremely important from a legal standing. It boils down to who actually owns the IP and assumes at this stage that there is no other individual or other party owning or laying claim to any part of the IP. If you created the idea, you own it and work for yourself, but are not a limited company, your name and address would go on the agreement. If you have your own business as a limited company and you want the IP to be owned by the company then your company name and address (and often the company registration number) would go on the agreement.

If you are not sure whether you or your company will own the IP, then it is quite normal for you to enter into two confidentiality agreements with the other party; one for you as an individual and one for your company. This is indeed very much a thumbnail sketch as to what name and address goes on the agreement and more detailed information on the complexity of IP ownership including relating to employers, employees and students can be found in Chapter Six;, "Who owns the idea? – Tiptoeing through the minefield". Please read this before deciding whose name and address goes on the agreement.

Period of the agreement

This differs from organisation to organisation and there are no hard, fast rules as to the period of the agreement; each party will have its own particular preference. The following will give you some idea as to why these periods are different.

Long Term
There are those agreements that once entered into are not limited by time and therefore continue in perpetuity. The main reason for this is that whilst your idea may only have a natural lifespan of two, five or even ten years, the other party, may well reveal their expertise and knowledge in developing your idea. Particularly in the larger organisations they would wish to ensure that this expertise and knowledge never comes into the public domain.

Medium Term
Five years is typical for many confidentiality agreements and once the agreement has expired, it is assumed that both parties have moved on significantly in that time and there is no risk to either party in revealing any proprietary information after that time.

Short Term
If your idea has a patent application filed, then a confidentiality agreement may only be in force for a period of 18 months or two years. The reason for this is if you wish to maintain and develop your patent beyond the first twelve months of having it filed, it will be published by the UK Intellectual Property Office (UK IPO) 18 months after the original filing (please note that published doesn't mean

granted!). See Chapter Five; "Protecting Your Idea". It is therefore in the public domain for all to see, so the other party could argue that they would not be in breach of the original agreement if they talked openly about what is in your patent and if the agreement has expired, then they would be free so to do.

Falling outside of the agreement

Many organisations include "get out clauses" in their agreements. The reason for such clauses is so it does not restrict them from going about the course of their business for the following reasons:

If it can be shown that the idea,

- Was known to the organisation prior to entering the agreement

- Has for whatever reason become known in the public domain

- Was already being developed by the organisation prior to entering into the agreement

The organisation will then have the right to cancel the agreement. There may be other "get out clauses" but those above tend to be the main three.

It is the third of these that causes most concern amongst individuals, particularly those of the more paranoid disposition as they believe that this is the clause where the organisation can decline any interest in their idea. In doing so are then free to copy the idea without proving that they were already developing similar in the first place. It has happened, and ideas have been "ripped off" in this manner.

One such instance was a major domestic appliance manufacturer who did just that. The individual who owned the IP in this case took the manufacturer to court and was eventually awarded several millions of pounds.

These cases are the exception more than the rule and the reason why organisations include that clause is to protect their investment. If they are genuinely already working on something similar, they may have invested hundreds of thousands of pounds in research, design, tooling and marketing. To take on your idea would possibly mean having to undo everything they have done so far and whilst your idea may be similar and in some respects, be better, for commercial reasons they will reject it.

Within a year, you could find they have a similar product in the market place. The question will always arise within you; "did they copy my idea?" Or for sound financial reasons, were they already working on a similar idea and have decided to stick with that? You'll probably never know the answer, but to find out could cost tens or hundreds of thousands of pounds if you pursued it through legal channels! You would have to be in possession of some substantive evidence to pursue such action.

Size of the agreement

This varies tremendously from one A4 page, up to as many as ten. It could be considered typical that the size of a confidentiality agreement would be between two and four pages.

Where to obtain a confidentiality agreement

There are various sources from which to obtain a confidentiality agreement for your own use. If you have a solicitor or an accountant, they may well have one you could use. There are other types of confidentiality agreements for different aspects of commerce, so make sure that it is a confidentiality agreement that covers disclosure of intellectual property. There are numerous websites that offer confidentiality agreements for sale and prices start from £8-£9 for a document. One useful source for an agreement is the UK IPO. A free download to print out can be obtained at the following link:

http://www.ipo.gov.uk/p-should-otherprotect-cda.htm

Chapter Three

Is your idea already out there? The point of truth!

No pain, no gain!

So your idea is probably mapped out, written up and recorded with a date stamp and where a working model is required to prove the basic principle, it's done and dusted! So far so good, but now comes the crunch time, finding out if your idea is out there already. This was referred to in Chapter One and the processes in this chapter will often be carried out in parallel with the drawing up and recording of your idea along with the working model build.

Finding out if your idea is already out there can be a painful experience, particularly if you find ten of them already in the market. Put bluntly, there is no gain without pain! If you find that there are ten other similar ideas out there already, you can find no significant differentiation between your idea and others and you cannot make or sell it significantly cheaper, then forget it! Yes, file your idea in the great idea wastebasket in the sky and move onto the next one! On the contrary, if by using thoroughly the processes in this chapter, you find that there are no or very few others in the market that may compete with your idea, then another stage is complete and your idea is shaping up nicely.

A warning here against self delusion! It is so easy to do these processes half-cock at this part of your idea development by not putting the in the required effort. In other words, if they are not undertaken thoroughly, you will delude yourself, but don't believe that in doing so you will pull the wool over the eyes of others, particularly as your idea progresses. Why does this delusion occur? Simply put, most people in their heart of hearts do not want to find out that their idea already exists and will undertake a very superficial search to convince themselves that their idea is the only one. From my experience, it is probably the Achilles heel of all individuals with ideas. Comments such as "It will sell in its thousands" or "my colleague has said there is nothing like it out there!" are on most occasions, hollow rhetoric. They have not done a search properly and subsequently against their claims, I have carried out a very basic search and proved that their claims are wrong – end of project!

You will have to be brutally honest with yourself in searching for competition to your idea and it will take quite a bit of work. Rest assured, if after all your diligent efforts you are still in the race, then your idea is moving forward quite nicely. Not only that, if you can show (and document) that you have done a thorough job in demonstrating that your idea is possibly original, novel and non-obvious, it carries enormous weight when convincing professionals who may help in developing your idea, particularly with those who may provide funding.

The three legged stool

The technical aspect of your idea may have already been proven through your working model, or you may still be working to develop it further. Whatever stage you are at with your working model, the next critical phase in the progression of your idea can be represented by the "three legged stool" model and satisfying the requirements of this model is of fundamental importance to succeeding with your idea.

IP
PROTECTION

MARKET

PRICE POINT

Your idea is the seat of the stool. Your idea now has to be supported by three important elements, each one a leg of the stool. If any one of those legs is weak or non-existent, then the stool, including the seat of your idea, topples over; it is no longer supported and does not function. Each leg is of vital importance to the seat of your idea. Each leg is represented by:

- IP Protection – can the idea be protected by way of patent, registered design or another form of IP protection? If it cannot, then barring exceptional circumstances, forget it! If you were thinking along the licensing route for your idea even at this early stage, then no company would wish to enter into a licence agreement. In nearly all circumstances, if you have no IP protection, you have no asset with which to negotiate with another party for the purpose of commercial exploitation.

- Market – If there is the possibility of protecting the IP, then the next step is to look at the market for your idea. You may believe that you have the best idea in the world and you may be able to possibly protect it, but if there is no market for it (nobody wants it), then stop there.

40

- Price Point – You've made sure that the first two legs are reasonably solid at this stage; you may be able to protect it and there appears to be a market for your idea. If however the market dictates that your idea will need to sell at around the £10 mark per unit and you think that you can only make yours for a retail price of £100 per unit, then failure is possibly just around the corner! There are though, occasional exceptions, an example being the integrated circuit which would replace many transistors in circuitry. Integrated circuits were incredibly expensive when they came to market, but the benefits of the solution far outweighed the cost.

Using various search criteria, we will look at these three "legs" in detail and demonstrate how you can satisfy yourself that your idea is cruising along nicely or that it's time to pack up and look for another idea.

Using search techniques

We will divide the searches into the three elements of the stool as outlined above.

IP protection
In Chapter One, reference was made to your creative asset (your intellectual property) and as to whether your asset will have any tangible value. The search process is important in determining whether there is any tangible value in such.

To find this out, searches of various databases, publications and use of relevant professional experts can determine whether your IP can possibly be protected. So what are we looking for? The objective is to find out:

- Whether your idea is already in the public domain

- If there are similar ideas to yours out there, is yours significantly different to the others for it to stand alone?

- What sort of protection would best cover your idea?

To be able to undertake the searches to their best effect, it is suggested that you read Chapter Five; "Protecting Your Idea" first.

In most cases, ideas will be practical and pragmatic; where there is function or a combination of functions that make the idea work. In these cases, it is most likely that the form of IP protection required would be that of a patent. Where a patent is not appropriate, a registered design may be the route to go if your idea is based around shape and look. If your idea is a function or a combination of functions and the look, shape and form has appeal which is important to how your idea is perceived, then both patent protection and a registered design may be necessary.

Two other forms to be considered are secrecy and know-how which are explained in Chapter Five; "Protecting Your Idea" as are copyright and trade mark which are, in nearly every case not relevant at this point. The searches undertaken for your idea's IP protection will be focussed on patents and registered designs using the relevant databases.

The first thing to do before commencing any searches is to draw up a list of keywords which relate to your idea. Take some time over this; you will gravitate towards only a few keywords as you are focussed on the sole objectives of your idea. To carry out a successful search, you will have to consider a far wider range of keywords outside the operational scope of your idea. There may be products out there being used for a completely different purpose, but using some of the functions in your idea. The important thing to remember here is that the fewer the keywords, the narrower the search results. It is almost impossible to draw up every conceivable keyword, but the greater the number, the less chance you'll miss something. This will demonstrate to others to whom you will show these results later on that you are taking a professional and diligent approach.

The following fictional example will give some idea as to how to draw up a list of keywords:

Mr. H comes up with a new idea that he believes has never been thought of before. He's got his own small business in the clothes manufacturing industry and for years he and his staff have been cutting fabric with knives. Mr. H has his eureka moment; a new cutting device that will do away with knives and improve the efficiency of his business. He comes up with something that he calls "a pair of shears".

He decides to do a database search to see if there is anything like his shears out there and comes up with the following keywords:

Shears, pair of shears, fabric cutting device, cloth cutting apparatus, material cutting device, large shears, hand held cutting device, two blade fabric cutting device
Not a bad start, and when he did his database search, he was pleased with his choice of keywords and was satisfied that he had found nothing like his idea. Of all the patents that he searched, none shared any of the functions or ways in which his idea worked. He believed nobody had invented a pair of shears before for fabric cutting and he was on his way to a fortune! Just think of all the clothes manufacturing businesses there are around the world; they'd all want some!

However, if he had considered other applications for his idea, then he would have been bitterly disappointed. So for what other applications could Mr. H's idea be used? If he had considered other industries that might use his device such as gardening, horticulture and medical for example and extended his keywords to include such applications, he could have come up with some addition keywords:

Gardening shears, cutting device for gardens, lawn cutting device, hand-held plant cutting device, grass border cutting apparatus, cutting device in surgery, medical cutting device, cutting apparatus in operations

In this extended search Mr. H came across a patent filed five years previous that was used in medical operations. It was a medical cutting device and looked remarkably similar to his, was hand held with two loops to place the fingers and thumb through to hold the device and had two blades that were pivoted and used as the cutting surface.

It meant that Mr. H might not be able to obtain a patent for his idea as it could be deemed that the know-how described in the medical device patent was in the public domain as it had been published. Also, if he pursued the idea of a patent, the medical equipment manufacturer could claim infringement of their patent. Although Mr. H had never seen the medical device before his search and he believed that he had his own original, creative idea, it could be argued that applying the cutting device to fabric would neither be novel nor non-obvious.

Mr. H has two choices; either to file a patent, then argue with the patent examiner or indeed the medical equipment manufacturer that his idea was a non-obvious use of the cutting device. The other choice would be to save a few thousand pounds by not pursuing the idea and therefore not filing a patent. Instead, he could buy his "fabric cutting shears" from the medical equipment manufacturer – it would still improve the efficiency of his business!

A patent attorney could argue the point,

"Mr H. could still be granted a patent by the examiner on the grounds that the examiner is happy that the use of the shears for cutting cloth was non-obvious to the medical equipment manufacturer."

There may be other ideas out there similar to yours that may not be direct competitors to your idea, yet may be used in different spheres of operation to yours. These other ideas may utilise some of the functions in your idea to make them work and it would not be apparent that this is the case unless you thoroughly searched for them with as many keywords as possible. Yes eventually you may be granted a patent, but there is a likelihood that it would be turned down.

The essential point to remember when doing a search is that patents that have been filed within the last 18 months of the day that you do your search will not show up on the patent databases as they have not been published. It will be important every so often during the development of your idea that you revisit your searches on the patent database to see if any new patents that have been subsequently published may conflict in any way with your idea.

Where to search
The first ports of call to find out whether your idea has mileage in terms of being novel and non-obvious are the patent databases. The UK IPO provides access to a free global database called Esp@cenet®. The database will allow you to use your keywords to search for other patents that are or may be similar to your idea. The database can be accessed at:

http://gb.espacenet.com/

This will take you to the first "search" page on the database.

There are different levels to searching for patents. To get started on your search, we will use the simplest search of all; click on "Quick Search" in the menu bar on the left. Keep the "Patent database" box as "worldwide". In the box "search terms", enter the first of your keywords.

A list of patents will be displayed with a title of one or two lines. They are not always in publication date order. Look at those that best match your keyword, then click on the title. The patent will then be displayed.

To delve even further into your search, there is another search that can be used and this will refer to specific classifications used by the UK IPO. Click back to the original page as shown above and click on "Classification Search". Under "Find Classification for Keywords", enter one or more of your keywords; then click the "Go" box. Look for the most appropriate description that matches your idea and then click on the box on the right hand side. You can click several of these boxes (up to four) if more than one description closely matches your idea. Once complete, click on "Copy" box and the classification numbers will be copied onto the "Advanced Search" page. Enter your keyword(s) in the "Keyword(s) in Title" box and then click on "Search".

Whichever the types of search you undertake, patents relating to the search process will then be displayed. The number displayed could be as few as two or three, but it could extend into the thousands; typing in the search word "umbrella" for example will bring up in excess of 16,000 search results!

Sometimes there will be no patents displayed. A word of caution, it doesn't necessarily mean that there are no patents published as you may have searched for example "car", yet patents could have been filed as "vehicle" or "automobile" or even "auto". Think of the alternative words for your keywords especially those that may be used by other countries such as the USA. You have to consider plurals; for example nothing may be found under cat eyes, but could be under cat's eyes or cats eyes.

So how many patents should you look at? In theory, every one that displays your keyword in the description (even if there are thousands), but in practice, at this early stage in your idea development, looking at the first 250 patents displayed will give a reasonably good indication as to which patents might be competing with your idea. Out of that 250, there may not be one single patent that competes with your idea (so long as you have included alternative words as described in the previous paragraph) and if that's the case, so far so good. You may be reasonably comfortable at this early stage that your idea may possibly have IP that has the potential to be protected. Conversely, there could be dozens of patents that look very similar to your idea. All is not lost at this stage, because professional assistance from a patent attorney later on may find ways around the problem of the competing patents, but the chances of having some IP that can be protected in your idea are not too high. See Chapter Five; "Protecting your idea".

If the patent search using your keywords throws up several patents that might compete, print off each of those and keep them for future reference. These will form part of your portfolio for your idea along with your list of keywords and will be used at a later stage of your idea development. Even if you cannot find one patent that may compete or conflict with your idea, print off half a dozen of the patents that were displayed in your patent search. This shows that you have been diligent in your patent searches and shows an understanding of the process undertaken.

So far, you have only undertaken a basic search and if there was no comparable search results using the words that you have, there may still be some hidden out there. There are more detailed searches that can be undertaken and these may well display other patents that your basic search has not revealed.

Search results displayed using the keyword "car"

How to differentiate between your idea and other patents
This really is a "practice makes perfect" exercise. What you are looking for is the common denominators in your idea and in those patent search results in the areas that compete or conflict.

48

When you click on the title of an individual patent, the full detail of the patent will be displayed. There are certain patents where no or very little detail will be displayed and where this is the case, there is little or no comparison that can be made between your idea and the patent displayed. The countries where often little or no detail is displayed in the patents are China, Russia, Taiwan and one or two of the lesser "patent active" countries.

The easiest place to start in comparing the patent displayed against your idea is the visual check by having a look at the drawing described as "mosaic". Click on the mosaic and have a look to see if there are any obvious conflicts with your idea. If it is difficult to ascertain whether there is anything obvious in the drawing that competes with your idea or even if there is something that you've spotted that may conflict, then click on "claims" and see if, by reading those claims, there is anything obvious that competes with your idea. Finally, if it is again difficult to ascertain whether there is anything obvious that may conflict with your idea, click on "description" and have a read through. This all takes time, but it is well worth doing.

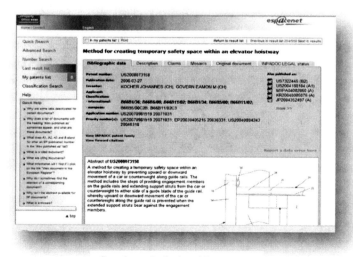

Search result displayed for a patent

By now you will have a reasonably good indication as to whether the "seat" of your idea is supported by a strong "leg" of IP. If you feel that this is the case, then you're on your way to building a robust "stool" and can move onto the next "leg" – the market. If there are serious doubts as to whether the IP in your idea can be protected, it is well worth arranging a meeting with a patent attorney. Patent attorneys will normally give you 20-30 minutes of their time free of charge to evaluate your idea and in that time, providing that you present them with your idea, keywords and search results, they will be able to give you a professional opinion as to whether the idea can possibly be protected. If in their opinion, it is highly unlikely that it can be protected, then the IP "leg" probably does not exist, or at best is extremely weak, so move on to the next idea!

You may ask the question, why shouldn't I go to a patent attorney in the first place before I've done any of the patent searches? Put simply, they may not be able to give a valued opinion in that first meeting without the evidence of your efforts. They may suggest that you do some more work on the idea or offer to do some of that work for you by way of a patent search. Whilst they may well do a better job than you could achieve, it will cost and a worldwide patent search could easily cost £1,500. It is still early days in the overall development of your idea and whilst I don't wish to dissuade you from using a patent attorney for the search, such costs may be better incurred further down the line.

To learn more on searching the patent databases, the British Library based at St. Pancras, London offers free in-house workshops at their Business & IP Centre which are normally one to two hours in length. Details of the workshops available along with dates can be found at:

www.bl.uk/bipc

The free courses that are available include "Introducing Patents Searching" and "Beginner's Guide to Intellectual Property" amongst other subjects.

If you are unable to get in to the British Library, their on-line distance e-learning courses are available at:

http://bipc-ecourses.bl.uk/

Once you have registered on-line, you can access the courses and having completed Course One "The Basics of Intellectual Property", you can move on to Course Two "Searching the Databases".

The British Library Business & IP Centre's On Line Courses

The market

There is an overlap here with the IP search insofar that the patents you've searched may possibly be in the market already. Searching the market is simpler than the IP searches, so you can breathe a sigh of relief here, but nonetheless, it is an exercise that's just as important. This search will also supplement your IP search on the basis that you might not have come across any similar IP on your patent search, but you may discover that there is an identical idea in the market that hasn't been filed as a patent application or a registered design. If this is the case, then it is unlikely you will be able to file for a patent for your idea as it is in the public domain already.

There are many ways of searching to assess whether there is a market for your idea. The most common are:

- Internet/Websites
- General Magazines
- Specialist Trade Publications
- Retail Stores
- Specialist Outlets
- Exhibitions and Trade Shows

Each of the above can be useful in trying to find what's out there in the market and sometimes you may have to call on all of them to further evaluate your idea. The line of least resistance is the use of the internet although some web sites only feature a limited range of their products on-line and if you relied solely on this source of reference, you could miss something. In these instances, it can be very useful to visit their stores or outlets to see if they have any products similar to your idea. Exhibitions and tradeshows are sometimes very useful as they often display new products that will not be launched until the following year; products that you would otherwise not find out about.

A lot of the process detailed in this chapter assumes that you have access to the internet. Using a search engine, enter the keywords that you used in the patent search, but omitting the words "apparatus" or "device". Don't just search UK sites; it is important to look at the whole worldwide web. Your idea may appear in markets overseas in which you are not remotely interested, but again it would prove that unfortunately, the idea is out there in the public domain.

In the patent searches, it was important to use as many keywords as possible including those that were outside the sphere of operation of your original idea (the "shears" for example), but initially in your market search, only use the keywords that are specific to the objective of your idea and the problem it intends to solve as this is the primary market for your idea and it is important to concentrate on this. These results will show whether, by finding similar products, there is a market for your idea.

Search results

Nothing like it in the market
Don't get too excited if you cannot find anything remotely like your idea in the market, it doesn't mean the world's your oyster and global domination will ensue! Rarely does something completely different, novel and brand new succeed in the market and if it does, it normally takes many years to do so. Remember from earlier on, a similar product to your idea may not be out there because it may have been tried and failed miserably.

Amongst the reasons why something completely different, novel and brand new won't succeed is:

- Manufacturers may have to change their tooling significantly

- Distributors and outlets have to change their marketing and distribution channels to achieve sales

- The inertia of the public/end users; a reluctance to change from what they know and use

If there is nothing like your idea in the market, trying to persuade manufacturers or distributors to licence the idea could be difficult for the reasons mentioned above. The cost for them to enter the market with your idea could be of such significance, that the rewards generated by sales would not justify the investment required.

Many examples in the market
On the other side of the coin, if there are many examples of your idea in the market, it is not a reason to feel despondent; this could be the proof you need that there is a market for such. So long as your idea has significant advantages / differences to those similar products and those advantages / differences are both novel and non-obvious, then this may be sufficient for you to obtain a form of IP protection, so presenting a market opportunity for you and your idea. The reasons for this are:

- The market is already established (maybe long established)

- Manufacturers and distributors have obvious distribution channels

- The market is of a significant size

- Future potential investors and licensees will be able to draw on established market statistics for proof of market and opportunity

- The public/end users are already familiar with such products

It is easier to enter such a market and obtain a market share although if there are a wide range of similar products out there, the market share your idea could gain may not be large, but could be significant enough to be commercially rewarding.

Just a few in the market
In addition to the two extremes of the market search detailed above, your search may only throw up two or three other products in the market. If this is the case, it would normally be:

- A specialist market where only a small number of products are sold per year

- A truly massive market dominated by two or three major global manufacturers

If your idea falls into the small specialist market, then there may again be a commercial opportunity here, but with numbers small, the rewards will not be tremendous, but they could help with the odd holiday or topping up the pension! If your idea falls into the massive market with two or three major "players" dominating, then the opportunity could be enormous for your idea. The problem here is the time that it takes to get the idea evaluated and into production; many companies in this field are planning way ahead, allocating budgets to projects and ideas sometimes up to three years ahead.

55

Basic market statistics

This is very important to your idea and your case for the need in the market. You will have to demonstrate both to yourself and others that your idea is a "must have need" and not a "might like option". There are many thumbnail statistics available on the internet not just demonstrating market size, but also the need. The demand for certain types of product is driven by legislation and if it can be demonstrated that your idea fulfils a need either through current or future legislation, then so much the better. Another useful source of statistics is reference libraries. The librarians are extremely knowledgeable and can offer tremendous assistance in sourcing relevant statistical data for your idea.

If you are able to get into the British Library's Business and IP Centre in London, there are some 30 high-value subscription databases (which are available for free) giving up-to-the minute company, business and industry information and financial news, such as Amadeus, Economist Intelligence Unit, ViewsWire, Euromonitor, Factiva, FAME, Financial Times, OneSource and the Complete Business Reference Adviser (COBRA).

The British Library team are most helpful in showing you how to use the databases and finding the market and statistical information that you need.

To look at how to go about sourcing market statistics, we'll look at an example.

Say your idea was a new type of smoke alarm for the domestic market which needs neither batteries nor mains wiring and never needed replacing.

By using the internet, the market size and statistics regarding use of smoke alarms can be obtained. This information could also be obtained from a reference library.

The aim is to give a "rounded view" of the market and the need for your idea. The statistics that you draw on will help to form a convincing argument as to why your idea is needed, where the demand is for such a concept and why it could be commercially viable.

The statistics to look for in this example would be:

- The number of dwellings in the UK

- The number of smoke alarms in use

- Accidents in the home relating to fires

- Legislation relating to fire detection

On the internet, search for:
"UK housing statistics"

Result:
www.communities.gov.uk

26,194,000 dwellings in the UK

"UK smoke alarm Statistics"
www.home-security-action.co.uk

4 million homes do not have smoke alarms.
6 million homes have alarms that are disabled.
40% of smoke alarms in the home do not work due to flat or missing batteries.
You are twice as likely to die in a fire if you do not have one.

"Fire statistics" www.fire.org.uk

In 2001/2002, there were
68,300 dwelling fires and
430 accidental deaths

"building regs smoke alarms" www.kiddefyrnetics.co.uk

In new build, you must use
AC hard wired only

So what can be gleaned from this information?

- 4 million homes do not have smoke alarms

- This means around 22 million homes do have a smoke alarm, but 40% of these 22 million homes (8.8 million) have alarms that do not work due to missing or flat batteries

- 6 million homes have alarms that are disabled

Putting it into a very simple overview, it could be considered that there is an available market of 4 million homes plus 8.8 million homes (a total of 12.8 million) that require a smoke alarm that neither needs batteries nor mains power and therefore doesn't need replacing. It is not clear in these statistics whether the 6 million disabled alarms are included in the 12.8 million, but it is assumed that this is the case. It also shows that in new build properties, there is no place for this new smoke alarm as alarms have to be hard wired. This market has been discounted completely and would be one to ignore.

Please be aware, that these statistics are only a "proof of principle" of market potential and detailed analysis of the true market potential of your idea will come much later (see

Chapter Eleven; "The big picture – Assessing the true market potential") when your idea has become a product nearing commercial exploitation. It is not necessary to undertake an extensive analysis at this stage as it can come at a cost and the processes in this book are about keeping costs to a minimum whilst your idea develops.

What this does not tell us is what share of that market could be taken by the new smoke alarm or how to go about obtaining it, but it does prove that there is an available market. At this stage, there is sufficient argument for the "market" leg of the "stool", but how will the new smoke alarm compare on the "price point" leg?

Price point

You have now looked at the market and found that there may be several products like your idea already selling and the statistics you've drawn on show that there is an available market. Now can it compete on price with the other products already selling? Bearing in mind at this stage, you only have a working model or one that is still being developed.

A simple rule of thumb is to assume that in the most part, your idea would be made of similar materials and be of similar size to those already in the market; it will have to be to compete. What is important to consider is what attributes does your idea have that sets it apart from the others? Do these attributes manifest themselves in just a different way to the other products or are they additions in your idea that are not present in the other products? If they are done in just a different way, then it can be reasonably assumed that your idea could sell for a similar price, but if they are additions, then the price of your idea may have to be higher than the other products.

It is difficult to give hard guidelines in respect of competitive price point comparisons as it really depends on what your idea is, its technical complexity along with that of other products in the market that compete with your idea. It will become instinctive after having done various comparisons and having studied how many components your idea has against the components in the competition, to know whether your idea can compete on price point. What you have to assume for now is that your idea would eventually be manufactured in similar numbers to those in the market; it would have to be to compete. It does not matter what the actual selling numbers are of the other products at this point, it is just an assumption that you have to make. Actual price point evaluation and use of price matrices to calculate the true price of your idea is addressed in Chapter Eleven; "The big picture – Assessing the true market potential".

An important point to take on board here is that whatever the price point is for your idea in the market, it will have to be manufactured for around a quarter of that price (the factory gate price). This is the price at which a product is manufactured and ready to leave the manufacturer. This of course is the reason why many products are manufactured in the Far East to hit that factory gate price.

A very easy mistake to make is only to take the manufacturing costs of your idea into consideration. For example many products have packaging, whether expanded polystyrene, blister packaging, cardboard plus any other variety of materials that may be needed to transport it, store it, display it and eventually keep it. These costs have to be taken into consideration in the eventual price point calculation. Other competing products will have taken that into consideration in their pricing, so if your idea is similar in materials and functionality to those products, then it is reasonable to assume that your idea will eventually be packaged for a similar price.

If, having done your work, you feel justified that your idea can sell at a price point the market will stand, the final leg of the "stool" is in place. If you are unsure whether your idea can compete or are having trouble in assessing what the price point may be, then don't give up at this stage; this is where a consultation with someone expert in their field could really help. This expert could be a designer, engineer or technician amongst others, but of course who you use depends on your idea and the way it has to be made. Within a few minutes of looking at your idea, generally they will be able to give you a rough idea as to how much it might cost to make based on commercial volumes of sales.

So the three legs appear firm, it looks like the IP in your idea may be able to be protected, there is an apparent market for it and it looks like your idea can compete on price with the competition; three robust legs. The "seat" is now looking pretty secure!

Make sure that you have printed off or photocopied all the evidence that supports the work that you have done in this and add it to your portfolio along with the search results; it could be extremely important at later stages when evidence is required to support your claims. Not only that, but it shows that you know what you're doing, you're being diligent at every stage and believe me, it will definitely help to impress the various bodies who you may need to get on board to help further develop your idea.

Chapter Four

Proof of principle – The working model

The importance of a working model

In Chapter One; "Setting Out", mention was made of turning your idea from a 2D notion into a 3D working model. It is of course the next progressive step and it is worth being reminded of the key points:

- It doesn't have to be aesthetic and "look good"

- It does not have to be made of the materials in which the final product would be manufactured

- Size does not matter within reason – so long as the model works; it's proved its point. If the market says it's got to be the size of a mouse and yours is the size of an elephant, you might have a problem convincing people!

- Don't over-engineer it with bells and whistles – you're trying to prove the key elements of your idea, that's all

These points have to be fully considered when thinking about building a working model. Of course you may be very capable of building such a model yourself and if that's the case, fine, but technical obstacles may present themselves. These could deter you from carrying on and either putting it on the back burner or abandoning it completely. If such obstacles are apparent, then do not be deterred, as this could be a good point at which to turn to a specialist in their field.

In using such specialists, you are utilising expertise that may well be far removed from your core knowledge or ability. Gaining that "third eye" could be invaluable, even if you are confident that you have made, in your opinion, a successful working model that proves the basic principles of your idea.

Technical obstacles – Using the specialists and gaining an invaluable third eye

Peter Ford, Head of Design at De Montfort University's Faculty of Art and Design works to a very basic rule:

"You need to try and see the end result early on"

Although very simplistic in its observation, IT IS TRUE AND VERY APT! Developing your idea from drawings, via a working model into a final prototype is only possible if you have some vision of the end result for your idea. This end result (how your idea will work and possibly how it will look) may change as your idea develops, but having the vision is important nonetheless.

Using expert, specialist help will be an essential resource; consulting a specialist will offer that invaluable "third eye" in the development of your idea. Universities are an excellent source of expertise and rest assured their input will be based on experience of designing and developing literally hundreds of products. Such a wealth of knowledge will help greatly, but it is only useful if you can convey your vision and objectives for that end result.

It is so easy to come up with an idea and it may be relatively easy to build your own working model, but without some sort of vision as to where it's going, it really is a pointless exercise. Peter Ford draws this analogy:

"It's a bit like having an idea for a book and writing a good beginning, but getting lost in the main body of the writing because you haven't considered the ending first, you don't know where you are going. In doing so, there will never be a successful conclusion if you can't 'see' that end result early on".

Yes how the book ends could change with little nuances, tweaks and surprises added here and there, but the overall clear objective to the conclusion will only have remained solid if the end result is clear from the outset. It is this vision of the end result that will steer you through that main body of development and help to keep you on track.

But as Peter Ford also says "you have to have the skill to completely change direction and re-group when you suddenly realise that in actual fact the original ending will not work and fundamentally has to change. This takes guts and stamina".

Even if you are an expert in the knowledge that encompasses your idea, and have succeeded in making a working model, to develop your idea further, the knowledge, expertise and unbiased advice from universities can often prove invaluable. Many universities are funded by local and central government to invest both time and money in innovative projects. Gerard Moran, Dean of the Faculty of Art and Design at De Montfort University explains:

"There are several funded initiatives that can be applied to new projects –Improving Business by Design and Resource Efficient Design (RED initiative) are two such examples that we have used to develop and prototype individuals' and small companies' ideas ready for commercialisation, supporting such development with a level of match funding."

Using such initiatives not only supplies the expertise to develop your idea into a working model but beyond that, into a manufacturer ready prototype. They can also offset a high percentage of the costs that you would otherwise incur in taking your project to such stages.

Further information on these initiatives can be found at:

www.dmudesign.ac.uk

Case study

Roger Derby travelled the length and breadth of the UK as a national sales manager, spending time in eateries on the motorways and major trunk roads; becoming increasingly fed up with food from salad bars or hot plates that would dry out or curl up as the day went along. Some outlets had covers over their salad bars and hotplates that solved this problem. It was however a bit of a juggling act holding a plate (particularly if hot!) in one hand, serving spoon in the other and opening a lid which sometimes did not stay open of its own free will.

Roger had an idea; what if the lids could be opened and closed without the need of a hand to do so using a button or switch of some sort? Surely this could solve several problems, amongst which those that Roger encountered; the drying out of food and the juggling act! He had certainly seen the end result early on.

He set about writing down his thoughts; what was technically needed to create such a device and how it would operate. From a design aspect, he considered a one-piece hinged lid that would open into the vertical position, then a lid that was like a shutter blind, sliding backwards and forwards and a lid of two halves that raised into a vertical position and closed again, creating a seal where the two halves met. He opted for the one-piece lid as he felt that it meant less moving parts than the blind and two-piece lid. In addition he believed it would be easier to clean and maintain.

After a couple of weeks of making notes and some basic drawings, Roger came up with a mechanism that seemed to solve the problem. The serving spoon would rest in a cradle and when lifted, would activate a switch (similar to that of a fuel nozzle at a petrol station) and in doing so, would drive an electric stepping motor which would raise a one-piece lid to the vertical position. Once the serving spoon is replaced in its cradle, this would reverse the motor to close the lid.

This seemed quite simple in theory, but would it actually work? Before incurring any costs in building a working model, Roger rightly so carried out a basic search on the Esp@cenet® database and could not find a published patent for a similar device. He also searched the internet, looking at catering equipment and any other types of dispensing, serving devices. Much to his relief, there was nothing similar that could be found; it was time to put theory into practice and make a working model.

Roger cobbled together several components he found in his garage:

- A flat wooden block

- A 9 volt electric stepping motor from a child's toy

- A button switch

- A 9 volt battery

- Various bits of wire

- A serving spoon

- A square piece of cardboard

- A home made circuit breaker

He assembled the components on the wooden block; wiring the switch to the motor via the battery and incorporating the circuit breaker in the circuit. Once the square piece of cardboard was attached to an extended shaft of the motor, he tested out his theory. The serving spoon was placed on the switch, holding it in the down position; he then connected the circuit to the battery. As he lifted the spoon off the switch, the circuit was connected; the motor rotated, raising the square piece of cardboard from the horizontal position to a vertical position and in doing so, the cardboard impacted on the home made circuit breaker, so breaking the circuit; the motor stopped. His lid was open!

In this example, he had incurred no costs in building this very basic working model, but Roger knew his technical limitations and this was as far as he could go. He had got the lid to open and though he didn't know how to reverse the process to close the lid he knew that it was only a matter of seeking the right help.

I have experienced this most satisfying of moments and seen it in others – a model that works! It is such an important step; what before was just a notion drawn on several pieces of paper now exists in glorious 3D and it functions! To see it actually do so is often quite exhilarating and provides that momentum to carry on.

The important points to remember in this example:

- Roger had a clear idea of what his end result would be

- The notes and the drawings had been accumulated over a few weeks to form a well rounded idea; not rushed into

- Roger had carried out his searches at an early stage. Although they are basic searches, it convinced him that there may be an opportunity for his idea

- A simple working model was built that demonstrated his idea in a practical fashion, although still very much in its formative stage, he showed that it could work

He was now stretching the boundaries of his knowledge, so what now? Before seeking expert help, Roger would be wise to review his working model and although it is very basic, it will be worth looking at the "what ifs?" – in other words, what if this or that fails? Through this scrutiny, he may well be able to improve that which has gone before. In addition, such an exercise will broaden his knowledge of his subject matter and equip him far better to answer questions and furnish more detailed information when seeking expert help.

Failure Mode and Effect Analysis (FMEA)

Failure Mode and Effect Analysis (FMEA) has been used for some sixty years in industry around the world and in its simplest terms, is implemented to assess what failures might occur and then mitigate or completely remove the consequences of failures in a design or manufacturing process. At its most fundamental, FMEA looks at the consequence of the failure of a component or several components and materials; it examines the effect on the overall function and what could possible result through such failure.

It is a practice that has allowed manufacturing to greatly improve and in certain circumstances has allowed back-up support systems to be implemented in a manufacturing process such that if "component A" fails, "component B" will step in and ensure that the task undertaken by "A" is sustained without any interruption or risk to the function of the product. To do so, component "B" has to be designed to perform its own function, but through FMEA, it was identified that it would require the function of component "A" incorporated within it. Without the proviso of component "B" incorporating the function of component "A", if "A" failed then the product and process would fail.

This may all seem very grand when considering your own idea, but using simple FMEA at this stage of your idea's development will prove important on two fronts:

- When employing expertise in developing your idea to a prototype stage, showing that you have considered these failure options could well save you a considerable amount of money in the long run instead of considering such effects after several thousands of pounds has been spent on a prototype. Yes the experts may well pick up on other FMEA issues, but your knowledge, having considered such, could prove important.

- Whether seeking a licensee for your idea or seeking finance, being well versed in FMEA relating to your idea and particularly if you have addressed all the perceived issues can only increase the viability of your idea and its becoming a success.

Roger Derby decided to look at FMEA in respect of his idea. Yes, he'd got a very basic working model but his thoughts had gone beyond that and spurred on by the success of his working model, created drawings of how he envisaged the salad bar or hot plate unit might look. The drawing of a

69

salad bar contained ten compartments; five either side of the unit, each containing a different type of salad. Each compartment would have a lid whose opening was activated by the lifting of a serving spoon out of its cradle by the side of the compartment. Using FMEA in its simplest of terms, Roger considered:

What if the serving spoon is

- Not replaced correctly by the user?

- Accidentally taken away by the user?

- Dropped on the floor and rendered unhygienic

What would then activate the switch to open the lid? Would the user then pull up the lid and possibly damage the mechanism?

If the cradle that holds the serving spoon:

- Became filled with sauces that were residual on the serving spoon when it was replaced in its cradle, could they subsequently leak through to the switch so shorting out the circuit that controls the opening mechanism?

- Accrued sauces off the spoon that dried out in the cradle, could the switch be "glued" in the off position?

Could the lid:

- Drop down of its own will, so injuring the fingers of the user?

Although having considered the issues of "what if this fails, what if that fails?" most of the solutions were outside the scope of Roger's ability, but he had at least taken them into account. He did however make one decision; the switch activated by a spoon was out of the equation as there were too many issues to address surrounding its use.

Roger worked on the basis that if the lid was of a lightweight material, then it would not need a massive motor and if the lid dropped down of its own will, the pressure would be minimal so would not damage fingers.

The removal of the spoon and switch as an activation method were deliberated over for a couple of weeks, then Roger came up with a possible solution; what if the switch was replaced by an infra red sensor on the front of each compartment which detected the presence of someone standing in front of it and in doing so, initiated the opening of the lid? This was the solution – hygienically it was far better; the serving spoon itself could then be left in a cradle within the compartment meaning only flat surfaces to clean on the outside of the salad bar unit.

It was the first cost that Roger incurred in getting his idea to the current stage; he bought an infra red sensor with a control circuit from an electronics store. Using the instructions supplied with the sensor, he set it up on his original rig on the wooden block and yes it worked. Roger walked up to the rig and when he was 9cms away, the sensor activated the electric motor and lifted the cardboard square.

Okay, the working model didn't look aesthetic, it wasn't made of the same materials as a production model, it was basic and there were no bells and whistles on it, but it worked!

What else can your idea do? Ripping up the rule book

On nearly every occasion, an individual or a company think only of one use for their idea – that for which the idea was created. In nearly every instance, there will be alternative uses for the idea, often far removed from the original use. It is important to look at these and rip up the rule book and think of even the most obscure uses for your idea – you will be surprised what you will come up with! Instantly, you can identify new markets for your idea that were never considered before. In my experience, it has been extremely rare for an idea to address one use only. Importantly, when filing for a patent, these alternative uses should be considered and included where possible in the patent application which we will come onto later.

Before continuing, take a bit of time here to make a list of what you think could be some alternative uses for Roger's idea.

Looking back at this case study, a key decision was made by going through the FMEA process to change the activation device from a switch to an infra red sensing device. Roger wondered if by using an infra red sensing device there could now be any other uses for his idea. He pondered for a couple of weeks as to what other uses his idea could be put and considered the main points of functionality of his idea:

- A presence sensing device

- The presence of someone within a defined distance of the device activating the raising of a lid on a container

- Food and other consumables were prevented from drying out

- Waste on consumables would be reduced and last longer

- Could be battery or mains powered

These points of functionality will help to form the claims part of a patent application. Yes they are at the most basic level, but a useful exercise nonetheless for the future.

With these points of functionality in mind and applying them to other ideas, Roger drew up his list:

- Automated cat or dog bowl – the lid only opening when the cat or dog came within a defined distance of the bowl. This would reduce the smell of cat or dog food in for example a kitchen and possibly reduce wastage

- An automated ice bucket for pubs – this would reduce the melting of ice and provide a useful point of sale area on the inside of the lid as it opens

- A back bar dispenser unit for, lemon and orange slices, cherries etc – could dramatically reduce wastage on sliced lemons and oranges. Roger thought that this doesn't sound like a great saving, but when he considered if using the automated lid device saved the waste of one lemon a day at 40p, then over 365 days it's a saving of £140 per year. This certainly made it worthwhile considering.

- A humidor – a what? Although he was starting to stretch his imagination, why not use it for a humidor? It could be used to contain cigars and only opens when the humidor is presented to someone who wishes to choose a cigar. Probably not Roger's most politically correct idea these days, but it was still worth thinking about.

Input – the think tank

As Roger had exhausted just about every possibility of other uses for his idea, he decided it would be a good idea to draw in one or two others for their input for suggestions for alternative uses (under confidentiality of course!). If you do so with yours, make sure that you have a confidentiality agreement in place – see Chapter Two; "Confidentiality-Keeping your idea out of the public domain" even if those to whom you are showing the idea are close friends. This is not to say that you don't trust them, but explain that you do this with everyone to whom you divulge your idea. Very rarely will anyone take offence with this request and if they do, is it really worth showing them your idea in the first place?

A couple of points to remember:

- Make sure that they know it is your idea and that by their input they are not laying any claim to your intellectual property

- That they will maintain confidentiality at all times. If they do not, then your idea is technically in the public domain and could forfeit the opportunity of you filing a patent application

Focussing on the opportunities

Those ideas that turn into commercial successes are "must have" ideas, and very rarely is success achieved with a "might like" concept – in the latter, the end user has more of a choice as to whether they would need to buy such an idea, often with the negatives outweighing the positives.

To turn an idea into a "must have" you need to identify those drivers that convince the end user of their need for such. Those drivers could be identified thus:

LEGISLATION - Is current or forthcoming legislation such that your idea addresses this legislation, making current products similar to your idea obsolete or expensive to upgrade to meet this legislation?

COST – can your idea be made to the equivalent standard of current products in terms of function, yet be sold at a price far lower in the market?

LIFECYCLE – does your idea have the potential to last far longer than equivalent or similar products?

NEED – does your idea address specific needs neither found nor identified in similar or equivalent products?

MARKET – are there specific emerging trends in the market that would greatly enhance your ideas potential?

Bearing these in mind, a useful exercise is to mark each of the uses for your idea against the drivers shown giving a score of 1-10, with 1 being "no" and 10 being "absolutely meets criteria of the driver". The top scoring use for your idea is the most likely "must have need", particularly if that score is in excess of 40.

This is not a finite exercise in making a decision as to whether you should opt for one opportunity or another, so guaranteeing success, but it will narrow down the options in a practical fashion and remove those "might likes" that have little or no chance of succeeding.

Proof of principle checklist

1   Working model – are you able to make this yourself? Irrespective of the answer, consult an expert in their field. Universities are an ideal place to start and an initial consultation is free. This will help focus your thoughts and will offer invaluable advice at this early stage in your idea's development

2   See the end result early on, not of the final appearance, but of the functionality to serve the need

3   Undertake an FMEA in detail. This may only be required at the simplest of levels, depending on your idea, but it could change aspects of your idea that open up other opportunities for such

4   List the main aspects of functionality. This will help in assessing other uses for your idea. It is also an early stage focussing on the possible claims that may go into your patent application

5   Look at all the other possible uses. Rip up the rule book and think laterally. At this stage, there are no "bad" uses for your idea

6   Consult one or two others for their thoughts on what other uses may be possible for your idea. *Do this under a confidentiality agreement*

7   Focus on the opportunities presented and discard the "might likes" leaving the "must haves". If you cannot find any "must have" uses, it may be time to discard your idea. Quite simply the costs that you

incur may never be or only just be recovered in pursuing a "might like" option

You are now in a position to look at the forms of protection available for your idea. This is not to say that you will file for IP protection at this stage, but consideration of such at this point is an important part of crystallising an idea into a possible tangible asset.

ABOVE ALL, IN THE PROCESSES DETAILED, TAKE YOUR TIME - DON'T RUSH AS YOU MAY MISS SOMETHING POTENTIALLY IMPORTANT TO THE FUTURE SUCCESS OF YOUR IDEA

Chapter Five

Protecting Your Idea

Why protect your idea?

If the idea is well protected, then should others copy and try to make commercial gain by doing so, you can always try to stop them. A form of IP protection for your idea increases your chances of doing so. Conversely if someone claims that you are infringing their idea and so affecting their business, you again have a legal standing to contest their claim.

Let's make it clear that having a form of IP protection for your idea does not guarantee success in either scenario, but generally, it certainly improves your chances. If ever such disputes lead to a court case, not only can the costs be horrendous, but such is the law that the arguments on either side can be very subjective and just as one judge may decide one way, another may make a very different decision.

Choosing the right form of protection

Take time to analyse your idea and work out which forms of IP protection are appropriate. Steve van Dulken in the British Library Research Office comments,

"Maybe only one form of IP protection will be appropriate, but there will be times when several forms of protection could be needed for your idea."

Patent
If your idea works in such a way that it has function or several functions which are novel and non-obvious, then a patent might be the appropriate form of IP protection.

Registered Design
Not all ideas have appeal in their shape or form, particularly aesthetic appeal that set them aside from others. If your idea does have such appeal, certainly if it offers you a distinct commercial advantage because of its appearance, then filing for a registered design could well be the appropriate route to take.

Trade Mark
Trade Marks are a useful form of protection, particularly if you have given your idea a name and created a logo to represent the idea. If the name and/or the logo help to give your idea an identity and in time reputation in the market, a trade mark may well be a good option.

Copyright
Copyright covers how you present an idea, but not the idea itself. It is an automatic right once the work is created. Copyright covers the written word, photographs, music, films, television and art. Copyright may not be applicable to your idea if it is a functional idea, but if for example you produce promotional material for your idea or an instruction manual on how to operate your idea, such material would be your copyright.

Secrecy and know-how
Occasionally, secrecy and know-how can be useful. If for example, your idea is a recipe or an algorithm embedded in a chip that only you and you alone know how to make, keeping it secret could be a distinct commercial advantage. If it became known, then others could copy. Whilst it is kept secret, and assuming there are commercial possibilities with your idea, you have an advantage.

Know-how can be described as a method of doing something that requires your skills and knowledge. Without such knowledge, the method or task could not be undertaken. If you have an idea for which you have filed a patent and you wish to licence the idea to a licensee, that licensee may well require your expertise or know-how in further developing the idea. This is often the case with academics in universities who have invented something for which they have filed a patent. Such may be their specialist knowledge and skills that the only way for a licensee to develop the invention to the point of commercial exploitation is to contractually engage the academic's know-how to do so.

Advantages of protecting your IP

This is dependent on which form of IP protection you use, but there are distinct advantages to owning correctly filed IP protection, the main one being a recognised legal ownership of the idea from the date on which the protection was filed or in the case of copyright, when it was created in written or recorded form through any visual, written, musical or artistic format

In respect of a well filed patent, one of the biggest advantages is if it can be shown through the early searches previously undertaken that it is possibly novel, non-obvious, has market potential and can be made and sold at the right price point, then it has a potential asset value. This asset value will be proven to a greater extent when further market analysis is undertaken. See Chapter Eleven; "The big picture – Assessing the true market potential". Of course, anything, whether it's a house or a car is only as valuable as to how much the market is prepared to pay for it, whether that is the end user in their multitudes or a licensee.

During your idea's development cycle, from that first notion through to commercial exploitation, the patent process may represent around 15% of the time and effort in the whole process, but represents possibly 75% of the total value in your idea, particularly when contemplating a licensing agreement. The reason for this is quite simple; demonstrating your idea in the form of a prototype may be the way in which you convince a manufacturer or distributor to want to licence your idea, but it is not the idea per se that they are licensing, it is the words, drawings and claims in your patent. This is the legal document that you are licensing or assigning to them. Just the same as that document gives you a legal standing, by licensing or assigning it gives them a legal standing in manufacturing and distributing the idea. You could have the best idea in the world, but without that patent, almost without exception, they would not wish to pursue it. This will be dealt with in detail in Chapter Thirteen; "Licensing, assigning and evaluation".

Design registration alone is not as black and white as patents in terms of advantages, but if its shape and form is appealing and this sets itself apart from other designs, then again this could have asset value. The asset value could be even greater if its appeal in the registered design is the appearance of an idea that also has a patent filed – a double whammy! The final design of an idea is normally towards the back end of the development cycle and therefore filing of such should be left until that point.

The advantage of a trade mark is that it gives your idea an identity; a name along with possibly a logo if you so choose. This can set it apart in terms of other products that may be similar to yours in the market. There are many famous brands that immediately identify a product and indeed in time, they become household names. However, to do so, there are many years of invention, design and marketing behind these products to establish them as such, not forgetting often at a cost of many millions pounds, dollars or euros.

If the name or logo that you have devised is representative of the idea itself, what it does, or why it exists, then often so much the better. The filing of a trade mark as with a registered design is best to consider towards the end of the development of your idea, when it's in prototype stage. At that point you are more likely to have a far better insight into its final form and to what the target market will be. This will shape your thought process as to what the idea might be called and what form the logo will take.

If you are going to manufacture your idea yourself, then a trademark could be of particular importance. It creates both an identity and reputation with your business.

If licensing your idea is your preference, then most likely, the manufacturer or distributor with whom you agree a licence will want to brand the idea themselves, incorporating it into a corporate brand name that has affinity with the stable of products that they already have. A note of caution here, I have come across two clients that were insisting the name that they had come up with for their idea should be the one used by the licensee. Net result was that one licence agreement was not concluded because of such insistence and the second licence agreement was concluded only after the owner of the idea backed off at the eleventh hour. Simply, in both cases, it was ego that interfered; both had thought that their name was infinitely superior to that which any licensee could possibly create.

Disadvantages of protecting your IP

Whilst advantages outweigh disadvantages, there are still disadvantages of which to be aware. No matter what type of protection you file for, in a relatively short period of time, it will be published by the UK IPO and other intellectual property offices if filing overseas; from then on, it's in the public domain for all to see.

Patent

Once you have filed for a patent, it will be published 18 months after your filing date and from then on, it will be available for all to see. Once this happens, everyone has access to all the published information. The downside here is that other parties who operate in a similar area to your idea could take elements of your idea and putting it bluntly, copy them and use them in their own products. Conversely, using their own "patent busters", they may try and work out how to use those elements of your idea and circumvent your patent without infringing any claims that you have made, possibly rendering any action that you might take against them (if you ever found out) fruitless.

There are companies who employ professionals to monitor new patent publications to see if there is the opportunity to do such things and the other side of the coin is they may well contact you to say that they feel you are infringing one of their patents.

It is worthy of note that if you never filed for a patent, then the disadvantages that are mentioned above could still occur.

Registered Design

Registering a design has similar disadvantages to those of patents; once published it is there for all to see. Always undertake a search of the registered design databases to see if your design stands in its own right. If you are sure that your particular idea has a very short market life, for example 6-12 months as have some toys as Christmas fads and you want to save costs, it may be better to use the element of surprise by launching in the market without a registered design. In doing so, by the time anyone tries to copy, the fad has passed. It is relevant to point out that if cost is not an issue and you timed the filing of your registered design so its publication coincides with the launch of your product, the element of surprise is still maintained.

Trade Mark

A decision on filing for a trade mark is somewhat dependent on how important it is for the perception of and the market potential for your idea. As with patents and registered designs, once published in this case in the trade marks register, it is there for public viewing. In addition, the UK IPO will notify those with similar marks, giving them a period of time to object to your mark and possibly giving grounds for refusal. In such circumstances, and particularly if your idea is a niche, low volume market and again, cost is an issue, it may be best not to file for a trade mark. Drawing such attention to your name and logo may not be such a good idea, but first before making that decision, you should always do a search on the trade mark database, so ensuring where possible, that you are not infringing any marks that are registered.

Part of these disadvantages can be mitigated if you seek professional advice for your IP protection from a patent attorney and together with their help, you can avoid as far as possible any infringement or passing-off issues, although this can never be guaranteed. If the IP protection you eventually file is professionally written and is robust, yes you run the risk of notifying other parties with similar IP protection, but on a positive note it can also signal a 'hands off' message.

How much will it cost? How long will it take?

The costs detailed below are in addition to those costs charged by a patent attorney to file your UK applications

Costs for international applications are not shown; information regarding such can be supplied by a patent attorney if it is appropriate for you to file internationally within 12 months of filing your UK application.

Patents
The total amount charged to process a UK patent application is £200. It is actually free to apply for the granting of a patent. The amount of £200 is broken down as follows:

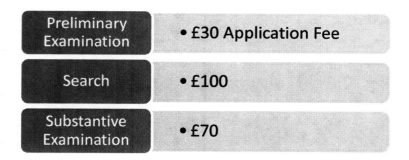

Preliminary Examination	• £30 Application Fee
Search	• £100
Substantive Examination	• £70

To ensure that your patent goes to publication, (subject to the UK IPO being satisfied with your application) the search (to see if your idea is novel and non-obvious) and the preliminary examination of your document have to be paid within 12 months of your filing date. Substantive examination (a full and detailed examination of your patent application) must be requested within six months of the publication of your patent. Patents can take up to three years to be granted with a maximum time limit of four and a half years.

Once a UK patent is granted, there are fees to be paid to renew it and keep it in force. This can be done for up to

20 years. The first renewal is on the fourth anniversary of your filing date and then every year after that. To view the current renewal fees, go to:

http://www.ipo.gov.uk/patent/p-manage/p-changerenew/p-changerenew-renew.htm

It is important to note that not all patents that are filed are published, particularly those where it would be against the national interest to do so or the device may be of a nature that could cause harm, for example armaments.

Trade Marks

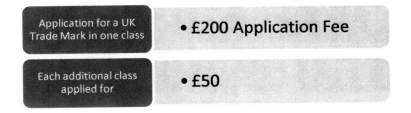

| Application for a UK Trade Mark in one class | • £200 Application Fee |
| Each additional class applied for | • £50 |

Two months after you received a filing receipt acknowledging your application, the UK IPO will complete an examination of your trade mark application and then send you an examination report either accepting or refusing your mark. If it is refused, the grounds on which it is will be explained. You then have to decide if you wish pursue an appeal procedure. If your mark is accepted, then after three weeks, letters are sent to other mark holders identified in the examination. Your mark is then published in the trade mark journal giving the right to anyone to object to your mark. If after three months and two weeks, the UK IPO receives no objections, then your trade mark registration certificate is issued.

You must renew your trade mark on the tenth anniversary of the original filing date and every 10 years after that. So long as you pay the renewals fees, there is no upper time limit as to how long you can keep the trade mark in force.

The renewal fees are:

£200 for the first or only class of the registration

£50 for each extra class

Further details on trade mark renewals can be found at:

http://www.ipo.gov.uk/tm/t-manage/t-changerenew/t-changerenew-renew.htm

In addition to the above, you may well find the Institute of Trade Mark Attorneys website a useful source of information, not only for trade marks, but for registered designs and copyright. More information can be found at:

http://www.itma.org.uk/

Registered Designs

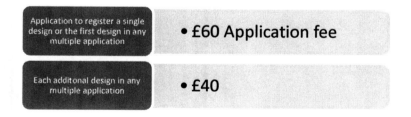

| Application to register a single design or the first design in any multiple application | • £60 Application fee |
| Each additonal design in any multiple application | • £40 |

You can if you so wish, because strategically it may suit you to do so, defer the publication of a design by up to 12 months. In this case, the fees are as follows:

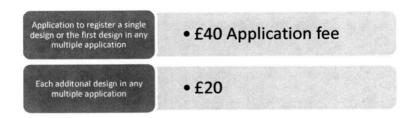

Application to register a single design or the first design in any multiple application	• £40 Application fee
Each additonal design in any multiple application	• £20

There are certain grounds for absolute refusal of a registered design application, but so long as your application falls outside of those, the examiner will not refuse your application and your design will be registered around three months after submitting your application.

A registered design can secure up to 25 years of monopoly, but it is a requirement if you wish to keep your registered design in force, to renew on the 5th anniversary of the registration date. Renewals follow every five years from then on. Further information on renewal fees can be found at:

http://www.ipo.gov.uk/design/d-manage/d-changerenew/d-changerenew-renew.htm

Challenge yourself; how should your idea be protected? Although in most cases, it is not necessary to protect your idea at this stage, it is worth planning what type of protection is or may be required for your idea. The key considerations are:

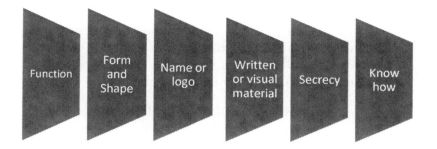

Function | Form and Shape | Name or logo | Written or visual material | Secrecy | Know how

It of course depends on the nature of your idea, but to give you a broad brush stroke as to what may be required in terms of these considerations, a couple of examples are given below:

- A synthetic roof tile made of recycled car tyres which will interlock with the tile below it to give a robust roof covering for domestic properties. The secret to its success was in the way the tyre remnants could be shredded and woven together at a microscopic level. This weave could then be moulded into a tile of any shape and form; malleable when being fitted, but incredibly strong and extremely long lasting. The shaping of the tile included a technique for creating and fixing device so that the tiles could be securely locked together. The owners of the IP (a small company) manufactured and distributed the product themselves, giving the product a name and instructions for use.

- A mathematical formula for a new number puzzle that was available on-line on the internet. This particular puzzle was new on the scene and could be calibrated to challenge those of the highest IQ, but its forte, discovered by accident, was its ability in assisting those who had incurred an illness or injury to the brain so helping them on the road to a full recovery. It was exalted as one of the best tools ever to be used in such cases. The owner of the mathematical formula had it securely embedded within his website for use on line and also offered a secure, downloadable version for computers. It was becoming well known by name alone.

Of course your idea will be different to those shown, but to give some ideas as to the thought processes that you should apply, it's important to see how these three ideas fit into a matrix along with the key considerations?

	function	form & shape	name or logo	written or visual material	Secrecy	Know How
Tile	√	√	√	√		
Puzzle			√	√	√	

The tile was prime patent application material because of the unique way in which it functioned. Its form and shape were appealing, so worthy of a registered design application. The name of the tile now had reputation, so a trade mark would be merited and copyright was automatically conferred on the instruction manual for use and the packaging material.

How the puzzle actually worked, the mathematics behind it, only the owner knew and there was no way that was going to be put into the public domain, so secrecy was maintained. The name of the game was well recognised, so a trade mark filing was becoming essential. The only way the puzzle could be used was by the instructions that came with it. These instructions had automatic copyright protection.

Always consider all aspects of IP protection when developing your idea. Those that are appropriate for your idea will not need to be filed all at the same time and perhaps only one form of protection might be appropriate, but still consider them all.

Financial planning for costs of IP protection

Looking at those two examples and then applying the relevant key considerations to your own idea, it is now time to look at the potential costs that you may incur in protecting your idea. This does not take into consideration protecting your idea worldwide or in specific countries as there will be time for this later. To start to plan costs that you may incur, the protection of your idea's intellectual property will be viewed from protecting it in the UK at this stage. A point to remember is by filing for a form of protection in the UK, should you decide to expand that protection to other parts of the world, then the date of your original filing in the UK is the date from which your filing is acknowledged by those other countries, called the priority date. That date takes priority of protection back to your original UK filing date and applies to patents, trade marks and registered designs. Copyright in the UK is an automatic international right.

In depth information on European and international IP protection can be found at:

http://www.ipo.gov.uk/abroad.htm

Patent, registered design, copyright and trade mark – Getting expert help

Why use their help?
There are different degrees of difficulty in filing for IP protection, the easiest of which is filing for a text only trade mark application, followed by a text and logo filing. This is followed by filing for a registered design and finally, the most complex of all is filing for a patent application.

In respect of all three, it is of course something that you can do yourself (trade mark and registered design applications are a case in hand), but a word of warning, with particular emphasis to patent applications; DON'T! Use a patent attorney to undertake the work on your behalf; this is by far the best route to take. Yes, this will incur costs but to a degree, it is dependent on how good your idea is and what ground work you have done from the previous chapters to prove its standing as to whether it is worth using their services and incurring those costs. If your idea is still holding fast then do so; if it isn't (and by now you should be aware if that is the case), then it is probably best to turn to your next idea instead.

As trade mark applications and registered design applications are generally more straightforward, the following focuses on patent applications.

The advantages of using a patent attorney

One distinct advantage of using a patent attorney is their experience in writing such applications; they have been doing it for years. A point to note: there are patent attorneys who are generalists and have a good working knowledge across many industry sectors and they may be more than capable of writing a satisfactory patent on your behalf. Always ask them what industry sectors they specialise in and preferably use one who has intimate knowledge in the sector in which your idea sits. This sometimes has to be traded off against what the sector specialists charge – a specialist in particular industry sectors can be more expensive than a generalist, so there is a balancing exercise. It is often best to go on recommendations from others as patent attorneys reputations are often built in that fashion.

You can find a list of patent attorneys in your area. Simply go to the Chartered Institute of Patent Attorneys website at:

http://www.cipa.org.uk/pages/home

and click on 'find a patent attorney' in the left hand menu bar

The advantages of using a patent attorney:

- They know the jargon and will write the patent in words relevant to such an application

- They may well consider technical aspects of your application that had passed you by. Indeed those aspects may appear to be minutiae, but they could turn out to be important.
- When using drawings in your application, a patent attorney's skill at reviewing the drawings and numbering them within the description and claims of your application is invaluable

- Claims – writing claims and prioritising them in the correct order is a consummate skill of a patent attorney. This is of great importance when, once your application has been filed, the UK IPO examine your patent and respond to the application. The success or failure of your application and any arguments that may ensue could well depend on these claims

- Having your patent attorney named as your 'agent' on your application shows that you have sought expert help and are serious about protecting your idea in a professional manner

- If there was ever a dispute over your patent either by you against another party or vice versa, your patent attorney is well positioned to argue on your behalf – after all he or she wrote your application and knows it inside out!

Patent attorneys – points to note

Patent attorneys are similar to solicitors or lawyers; they accept a brief and will in most cases, only act upon the information provided. This is not a criticism of their professional abilities, more an observation of their skill sets. They will often be able to quote from previous cases that they have dealt with and usefully adapt some of that knowledge and apply it to your "case". I have dealt with patent attorneys who have no commercial awareness whatsoever in respect of particular ideas that have been presented to them and an argument could be made here; do they need to have? In those cases, the commercial attributes of such ideas were not presented and yes they did write a satisfactory patent application, but to the exclusion of certain aspects within the ideas that could have made the applications broader and even more robust.

I cannot emphasis enough the ground work that you have to do prior to seeing a patent attorney, by carrying out extensive searches in the patent databases and the markets. This is another reason for having sought some expert advice on the development of your idea from relevant institutions such as universities. The more work carried out at this early stage along with the information gathered through your efforts will certainly stand you in good stead when an application is written.

Another point to bear in mind; within five minutes of a consultation with a patent attorney, they can see if you are a time waster or not. Providing them with all the information you can at that first consultation will get them on your side right away. They don't just wish to write a patent application for you, they want to be in there for the long run, whether arguing on your behalf with the UKIPO examiner or subsequently making additional filings in other countries. Who knows, you may provide them with further work on a trade mark application or a registered design application for your idea! Of course, this means forking out more money, but patent attorneys are no different from any other business — they're in it to create new business and ongoing relationships.

Cost examples of using a patent attorney's services are detailed in Chapter Nine; "Planning the development of your idea".

Chapter Six

Who owns the idea? Tip-toeing through the minefield!

Whenever a novel idea is created and developed, particularly a commercially viable one, the matter of who owns it must be addressed from the outset. This will become more important in the final stages of your idea's development, particularly when commercialising the idea. Whether you will start a new company to manufacture and sell it or use your existing company to do the same or alternatively, licence your idea to a major manufacturer or distributor, clarity of ownership is essential from several legal aspects. It is also a major issue in confidentiality agreements.

For trade mark and registered design applications, in terms of ownership their respective forms ask for the applicant details. This section is filled in with the details of who owns the mark or design as the applicant. A patent application form is slightly different; there is a requirement to fill in the applicant details, but there is also another box for "inventor". Before completing the sections of any of these forms, ownership has to be clear and transparent. If that is not the case, all sorts of complications and problems even costly ones, can arise. There is no absolute, finite answer and under certain circumstances, the issue of ownership can be a minefield. This chapter is designed to help you understand the issues, but ultimately, you, your colleagues and maybe companies or institutions with whom you may be involved will have to agree the ownership of the IP.

If using a patent attorney, all three forms require the "agent" box to be completed. That is exactly what is meant, the patent attorney is only acting as an agent and has no rights of ownership to the IP in the applications.

When an individual owns the IP

Often the easiest form of ownership, or so it seems is when one person or one company wholly owns the intellectual property, yet there are issues to consider here.

If you are:

- an individual who does not work for an employer

- not undertaking contracted work

- a sole trader

- the sole owner of a limited company

and you have come up with an idea which is acknowledged by others that it is your idea alone, then you (or if you choose your wholly owned limited company) would more than likely be the sole owner of the IP and the applicant that would be entered on the forms for filing.

In respect of a patent application, as an individual, you would be entered as "the inventor" and "the applicant". If your work on establishing the viability of your idea proves through the early processes in this book that it does have a reasonable chance of being commercially viable, then before filing an application, it is wise to talk to an accountant to see if say, for tax purposes, it might be worth setting up a limited company or if you do already wholly own your own limited company, for the company to own the intellectual property. You could be named as the inventor (in the case of patents) and the newly set up or existing company could be the applicant. The company would then be the legal owner of the intellectual property.

A word of caution, if you do this, you have relinquished ownership of your idea to your company. Whilst there are endless legal considerations in doing so, just two simple questions to ask yourself:

- What if I sell the company?

- What if the company goes bust?

If you sold your company, then part of the attraction by the other party in buying your business may be the commercial success of your idea and the fact that the company owns it. This would invariably be reflected in the price that they are prepared to pay and you may well be happy with this.

If your company goes bust or you have to make a forced sale because of the serious state of its finances, then as you no longer own the IP, you may possibly have waved goodbye to it for good.

Another point to think about is if from the outset you have owned the IP and filed the application with you as the applicant, you could then enter into a contract with your company that gives it exclusive rights to use your IP and you could have a clause written in that states you have the choice to make the contract null and void if the company is either eventually sold or goes out of business and that the use of the IP wholly reverts back to you. You would then be free then do what you wish with your IP.

There are many other scenarios that can arise but those above are amongst the most common to consider. Sound legal and/or financial advice should be sought at all times when contemplating the above or other possible scenarios.

When more than one individual is involved in the ownership of the IP

If from the outset, two or more individuals meeting the above criteria as in "When an individual owns the IP" or a partnership of two or more individuals own the IP, then on most occasions, the names of those that own it collectively are entered together as applicants on the forms. In respect of patents, all names are in many cases entered in the "inventors" box as well. Being named as the inventors is an acknowledgement that those individuals are the inventors, not the owners of the IP unless they are also named in the "applicants" box.

Attention has to be drawn to a point of ownership here and we will consider a patent application made by three individuals all named as applicants and inventors on the forms for filing. As mentioned previously, being named as inventors is an acknowledgement of exactly that; they are the three inventors. In the "applicant" box all three are named as well, making them joint owners of the intellectual property.

Note here, that this joint ownership alone does not mean that they have to jointly decide what to do with the intellectual property. Each of the joint owners may make use of, offer to sell or sell the patent without seeking permission from the other two. It's almost for want of a better term as used in banking, a "joint and several" ownership. In the case of this patent ownership, one, two or all three individual(s) can decide what to do with the intellectual property.

Another issue to be drawn to your attention; the three inventors also stated as applicants contributed in different amounts to the total sum of the intellectual property in proportions of 10%, 35% and 55%. This means that the inventor who contributed 10% could as an applicant make

99

use of, offer to sell or sell the whole of the IP without reference to the other two; not a bad deal is it?

The inventor who contributed 10% to the IP could argue with the other two, if it got to that stage, that without his 10% contribution, the remaining 90% of inventiveness would have been worthless; the IP in his opinion would have zero value.

Hopefully it is becoming clear as to how messy the issues could get.

To minimise the risk of such occurring, it is best where there are joint applicants to avoid such contentious issues and get a "heads of agreement" drawn up before filing an application that clearly defines what each of the owners can and can't do and as to how, if the IP is commercially successful, any revenue / income will be divided between the owners. There will usually be other terms included, but these are open to agreement between the individuals. An accountant or solicitor can draft this agreement between the individual owners of the IP, and yes this will cost, but maybe for £200-£300, it could turn out to be a very sound investment if early on in the development of your idea, there could appear to be commercial opportunities; later on, thousands or hundreds of thousands of pounds of revenue could be involved.

Companies' and Institutions' rights to IP created by an individual

There are various commercial agreements that are made between companies or institutions and individuals either working for, studying within or contracted to them and this section sets out some examples but is not limited to such. I have come across many examples where individuals have pursued some excellent ideas with great gusto and effort only never to have checked what their standing is should "their" idea be commercialised. We will look at some arrangements at the most basic of levels without going into legal jargon.

The simplest of all agreements which mainly relates to companies as opposed to institutions is that as part of their terms or contract of employment, there will be an agreement written in that any intellectual property created by the employee belongs wholly to the company and the employee has no rights whatsoever to any benefit from the development or commercialisation of the IP, although they would probably be named as the only or one of several inventors on the patent.

A point to note here is that some companies only relate this aspect of the agreement to hours that are worked within the company i.e. 37.5 hours per week; others will relate it the period of the terms or the contract of employment.

Here is the subtle difference; if the agreement relates to the hours worked within the company, then any IP that is created by the employee outside of these hours of employment, for example in the evenings or at the weekends could in most instances belong to the employee. There have been cases though where a company has argued that whilst the intellectual property was created and developed by the employee outside of their working hours, they could not have done so without the knowledge and know how of the company accrued during their working hours. There is no easy answer in this instance as it is a very subjective argument and one of which to be aware.

Where the agreement relates to the total period of the terms or the contract of employment, whether in working hours or not, this is much simpler – if the employee creates IP either within working hours or otherwise, this means in nearly all cases that the IP belongs to the company.

Some companies and institutions, depending on the nature of their business will have confidentiality and secrecy clauses that can automatically prevent an employee creating and developing any IP whatsoever, so mitigating the risk of bringing it into the public domain, even if by accident or innocently.

Surprisingly, there are thousands of companies that make no reference to the creation and development of intellectual property by an employee in their terms or contract of employment, either because it is not relevant to the nature of their business or they have never thought about it.

Where an individual is contracted as an external consultant or as a project manager to a company, similar agreements as above could be written in applying to the length of the contract. It is always worth checking this out if this is the case when undertaking externally contracted work.

An encouraging trend over the last few years relates to institutions and particularly universities; more and more not only have agreements which reward the individual, whether as an employee or student for creating and commercialising intellectual property, but have increased the level of the reward to encourage innovation and entrepreneurship. Each university is different in how they treat this – there is no institutional standard. Most will want to own the IP unless they otherwise put it in writing to release it to the individual or individuals. The main reason for wishing to own it is that most students and particularly staff of an academic background will have utilised all of their university's resources they can muster in creating, developing and testing the IP.

In its most basic format, the agreement that is put in place for the IP works on a stepped level of remuneration for the individual. Using an analogy, the income from the commercialisation of the IP is split very much between the

individual and the university in the way a wage earner
splits their income with the Revenue:

Income Bands (*theoretical*)	Individual	University
£1 - 15,000	100%	0%
£15,001 - £50,000	80%	20%
£50,001 - £100,000	70%	30%
£100,000 +	60%	40%

The income would normally be calculated as a royalty
income if a licence agreement is in place with another
party, or if a new spin-out business is created to
commercialise the IP, net income after any associated costs
are deducted and whilst this is a theoretical example as it
changes from institution to institution, it gives a feel for
what could be earned with commercially successful IP.

Whilst any of the above could apply to you and your idea, it
is well worth checking out with your employer (company or
institution) to see where you stand legally in possibly
creating, developing and commercialising your own idea
within the confines of their walls or indeed outside. It is an
often neglected, but an important aspect to consider and
you would be wise to do this at a very early stage, otherwise
it could later prove a complex hassle and possibly
expensive.

Ownership in confidentiality agreements

Before having filed for any form of protection, there will be
a requirement to disclose your idea under a confidentiality
agreement to those who can assist in furthering its
development. As your idea may still be at an early stage in
the development cycle (still on paper or maybe a basic
working model) you may feel that you need such help.

If you are an individual who exclusively owns the IP, then apart from the party with whom you are entering into confidentiality, there is only a need for your name and address to go on to the agreement. If there are two or more individuals who jointly own the IP then all names and addresses should go on to the agreement whether as individuals or as a partnership.

If you own a company and are not yet decided as to whether to file for protection in your name or the company, then you should enter into two confidentiality agreements with the other party; one on behalf of yourself and one on behalf of your company. As confidentiality agreements are usually two-way in terms of what they cover, the party to whom you are divulging your idea may insist that this is the case. This is quite normal practice.

If you only ever enter into a confidentiality agreement when you have filed for some form of IP protection, then normally the agreement is between the other party and the named applicant(s) on the filing.

It was mentioned towards the beginning of this book that a confidentiality agreement, although a legal document, is in effect the spirit of both or more parties maintaining total confidentiality - that neither party wishes to breach such confidentiality and covering every angle of that confidentiality in the form of the agreement or multiple agreements when required reduces the risk of legal loopholes.

What if applicants fall out?

Sadly on occasions, applicants (usually individuals) who are named on the filed application forms decide to go their own ways. This could be for a plethora of reasons and money is often one of the main causes. The costs to develop the idea may become far greater than anticipated and a difference of opinion occurs. Another reason that often crops up is disagreements on design and development strategy for the idea; one wants bells on it, another wants whistles; battery or mains powered etc? The combination of differences of opinion could be endless and often the design and development strategy is tied up with money.

One would hope that these differences can be resolved, but in a good number of cases, they can't. In such a situation, agreement has to be reached as to which way to go. If one applicant or more in some instances still wish to pursue the idea and the other or others do not, then subject to mutual agreement if that is possible, a transfer of ownership can be undertaken. The terms of any agreement between the applicants are not covered in this book and sound financial and legal advice should be sought from the appropriate professional body which in most instances are accountants or lawyers and yes, there will be a charge for this.

The transfer of ownership of the patent, trade mark or registered design can be processed via forms available from the UK IPO. It is free to change ownership of a patent and registered design through the UK IPO, but there is a charge of £50 to do so for a trade mark. You may wish this to be handled through a patent attorney acting as your agent and it is something that I would suggest. There is more detail together with downloadable forms at the following UK IPO pages:

Patents
http://www.ipo.gov.uk/patent/p-manage/p-changerenew/p-changerenew-assign.htm

Trade Marks
http://www.ipo.gov.uk/tm/t-manage/t-sell.htm

Registered Designs
http://www.ipo.gov.uk/design/d-manage/d-changerenew/d-changerenew-nameaddress.htm

Assigning IP to other parties

Assigning your IP to other parties is possible and the same process applies as detailed in the preceding chapter. If it is not for the reason above, it could be that (if owned by you or several of you) you are selling your business (limited company or partnership) in part or whole and the deal is that the purchaser acquires your filed or granted IP. Again I would suggest that this is handled through a patent attorney as your agent.

Another reason to consider assignment of your intellectual property is when you have approached a company or they may have approached you and instead of entering into a licence agreement, they wish to acquire your IP and all rights attached for a sum of money. This could be a lump sum up-front payment or several payments spread over a period of time. In nearly all cases, but I point out not on every occasion, a transfer of ownership of the IP would be part of the deal. In these cases, the transfer forms mentioned previously would apply.

In assigning IP to another party, warranties are often required. A warranty is where you guarantee to the assignee that the IP that you are assigning to them is protected as per the documentation that you have provided and that any appropriate renewal fees have been paid and up to date. The documentation you provide may well be enough to suffice, but they may require a warranty nonetheless.

If at any point, it is found by the assignee that the IP protection is not what it should be, by you entering into a warranty, they will have a legal right to seek financial compensation from you. In most cases, a warranty tends to be in force for twelve to eighteen months, after which time, they no longer have any legal recourse to you for financial compensation.

Chapter Seven

Protect your idea now or later? Filing protection for your intellectual property

Several chapters back, it was mentioned that protecting the intellectual property in your idea may represent around 15% of the effort in your idea's complete development, but it could be worth 75% of the value of the whole project, particularly when it comes to licensing your idea to another party; it is the IP protection that they are licensing, not the working model or prototype. The old caveat applies though in that the protection has to be robust, well filed and stand up to examination/inspection.

That caveat is compounded by an extremely important consideration. Your IP protection has to be filed as near as possible to the time in your idea's development cycle that will optimise the robustness of the protection to its fullest and maximise its commercial opportunity.

You may well ask at this juncture, when is *this* time in the development cycle that I should file for protection of the IP in my idea? A good question and there is no golden rule as to when it is the absolute correct time to file, but what follows will give you some very important guidelines as to when to do so. As all ideas that are developed have their own unique attributes, the time to do so will be different for each.

Put simply:

- File too early and you could render your protection worthless

- File too late and you could just miss the boat – someone else may have beaten you to it

The former of the two is, in my opinion, far more risky than the latter and I will explain the reasons behind this rationale.

Filing too early

The natural thing to consider when you believe that your idea is the next must have need that everyone will want is "I must protect it" and protect it fast in case someone copies it. I have been there with the very first patent application that I filed and what a learning curve that was! I filed it almost immediately after having done some basic drawings. It turned out that my patent application was worthless as I'd firstly filed far too early and secondly, the idea changed beyond recognition once I developed it further. I could no longer utilise most of the claims in my patent application for two reasons; technology had changed and some of the technology in the patent application was now outdated for the purpose it served and secondly, I discovered a far better way of getting my idea to work, still serving the purpose for which it was intended, but faster, smaller and cheaper.

My problem was that I could not file a new patent application as after 18 months of toying with the development of the idea, my first patent application was published and in the public domain. If I filed a new patent application, some of the claims would be the same as those in the first application. It would possibly be opposed by the examiner at the UK IPO as those claims were in the public domain. Another point is that if I did pursue a new filing, then I would have spent money on filing two patents applications in total. Weighing up the cost of a second patent application (bearing in mind it had to be completely rewritten for the best part) together with the likelihood that it would not be granted, it was decided to drop the idea altogether.

A patent application can be amended once filed, but there are strict rules governing such. You can remove elements of your application and this is of particular relevance when your patent is examined by the UKIPO and the result of the examination shows that there are other patents cited against your application that you may infringe. By taking out such elements, you may be able to remove the examiner's citations and your application can pass to the next stage without any objections from the examiner. Here lies the problem; in having removed them you may have destroyed the fundamental functions of your idea, so rendering the remaining functionality of your idea within your application worthless on its own.

You cannot add anything to a patent once it has been filed; for example during the development of your idea, you discover that by adding a "widget" or by replacing one with another, the functionality is vastly improved beyond recognition and results in a reduction by half of the manufacturing cost, then it is too late, you've missed the "widget boat"!

If that widget was a make or break situation for the success of your idea, you could if you so wish abandon the original patent application within 12 months of the filing. You could then file for a new patent with the widget included, but it will be a new filing date as the old filing date disappeared with the abandonment of the original patent.

Filing too late

This is an aspect that worries many – if they leave filing their application to a point late in their idea's development they believe there will be ideas the same as theirs in their multitudes that beat them to a filing date.

It is far better to get your idea to a point where the description along with the claims and if necessary drawings that you wish to put into your patent application are those that are representative of your idea nearing its final development stage. Yes there are exceptions to this – your idea may be so simple and its functionality obvious and robust at the drawing board stage; it is clear that it will not change. It may be best to file your patent application at that early stage, but this is the exception rather than the rule.

Another point to bear in mind – if another party filed for a patent application with the same idea as yours a day before you filed yours, you will not be able to see if this is the case until 18 months later when their patent is published and in the public domain. There is no point in waiting anxiously for 18 months to see if you've been beaten by a day, a week or even a month. It's far better to continue your idea's development and file at the appropriate time.

What do you decide to file for?

There are three aspects to look at here:

- Functionality
- Appearance
- Identity

Whilst digesting the contents of this section, it will be important to read Chapter Eight; "Planning the development of your idea".

This is an example of how many other processes can run concurrently; the progress of your idea could well develop simultaneously on several different fronts.

Functionality
Protection of your idea's IP in terms of functionality relates in nearly all cases to the filing of a patent application. You need to have finalised all aspects of your idea's function; how many elements that it has to its function along with how these elements work together. This could occur in exceptional circumstances when the idea is still at the drawing board stage, a little more likely at the working model stage and on most occasions, realised when in the prototype stage. Having looked at the patent databases, completed the basic market search and developed the idea to the prototype stage using the relevant expertise, you are more than likely ready to include all the functionality that you can foresee into a patent application.

Appearance
A lot of ideas that are created have purely functional elements to them, they require no smart shape, external design or a particular appearance to be successful; they are there to carry out a function and are often hidden away. For this reason, appearance is not important.

If the appearance of your idea, whether or not your idea contains functionality is important to its appeal, then getting this appearance right is important before filing for a registered design. Very rarely does this appearance materialise before a decent prototype is in existence. The prototype may not have every finite element of its appearance and shape, but you are well on your way to achieving this at the prototype stage. It may now be appropriate to file a registered design.

Do not file until you have:

- Built a working model
- Consulted an expert in their field
- Carried out IP searches
- Undertaken basic market analysis
- Seen a Patent Attorney
- Have a prototype in an advanced stage

Identity

Everything is shaping up well; the functionality of your idea seems to be working well, the appearance (if important to its appeal) is coming along nicely and now an identity to your idea seems appropriate, whether this is just for the purpose of a project name for your idea or if it is to be used in its own standing in the market as a brand. If functionality and appearance are concluding as planned, then using a name for your idea could be extremely useful at this stage, particularly if the name describes the idea. There is very little point in filing a trade mark application if you are intending to licence your idea as the licensee will wish to use their own name and identity for the product, but if you are thinking of making and selling it yourself, then filing a trade mark with the name, possibly along with a logo is not a bad idea at all. It is not essential at this stage and can be undertaken later but in doing so, now another hatch is battened down in terms of all round IP protection.

Tactical Advantage

As mentioned, timing of the filing for the protection of intellectual property is important, but there is an essential consideration to be taken into account, particularly with patent applications; as your patent is not published for 18 months after filing you have a "golden window" of covert activity. Once the 18 months is up after filing, your patent is there for the world to see but prior to this, assuming that you have confidentiality agreements in place with those with whom you have worked on your idea, no third party knows about it.

Tactically, this is a distinct commercial advantage, particularly in seeking a licence agreement and is even useful if you wish to set up your own business or incorporate your idea into an existing company. If the start of commercial exploitation can be achieved prior to the end of that golden window of covert activity, then the market is taken by surprise. If for whatever reason, you cannot exploit your idea say for 12 months after the end of that golden window, the rest of the world has had a year to look at your patent and play catch up or even beat you into the market with a similar product.

The emphasis here is again on the timing of the filing of your patent application; file at the beginning of the development cycle i.e. drawing board stage, then by the time your prototype is ready to be exploited commercially, there may be little time left before your patent is published. File at the prototype stage and you may well have over a year of covert activity to commercially exploit your idea – a distinct tactical advantage.

One misconception held by many is that you have to have a patent granted before a company will consider licensing your idea; absolutely not – a patent may take up to a maximum permitted time of four and a half years to be granted and by that time, your idea has been in the public domain for nearly three years! Every licence agreement that I have put in place was concluded during the filing stage of a patent prior to publication – that "golden window".

Filing later in the development cycle with a robust, professionally written patent has distinct appeal to possible licensees, allowing you to pass over the baton of opportunity for them to have time to further develop and then exploit your idea during that time Of course, a licensee has to be bowled over with the idea in the first place, but having this tactical advantage of time can only add to its appeal.

Chapter Eight

Planning the development of your idea

It is time now to consolidate your position in terms of where your idea is at and where it is headed, with the three main considerations being:

- Cost

- Time

- Expertise

So far, you will have undertaken basic searches that seem to show that your idea is potentially viable and will have worked it up from sketches into feasible drawings leading to a first stage working model. All this has been at minimal or even zero cost, only your time and effort has taken your idea to where it is now.

You have arrived at a fork in the road; you can take the left fork and continue developing your idea yourself, making what you perceive to be a satisfactory prototype, writing and filing your own patent application and if relevant, applying for a registered design or even a trade mark. You could take the right fork and seek out all the relevant professional help and expertise available for your idea.

Minimising risk

There is an element of a trade-off now between cost, time and expertise; so what is the best option? Without doubt, employing the services of a patent attorney to write and file on your behalf the intellectual property protection is paramount and minimises the risk of filing an inadequate or in some cases a woefully bad application, particularly in respect of patents. I have yet to see a good, robust patent written by an individual other than a patent attorney.

Using the relevant expertise to build a prototype is in most cases essential to further prove that your idea works. Using this expertise, particularly from the point of FMEA is invaluable. Whilst you will have carried out your own FMEA at the drawing and working model stage, there will be issues that you could never discover yourself; they will arise when use of materials and added functionality are applied to the first stage prototype through to final prototype sign-off. Always expect the unexpected at this stage of the development and relevant expertise is essential.

Using all this professional help minimises the risk of anything going wrong and presents you with a consummate package of product and protection, but at a cost.

Doing it yourself

With all the ideas that I have seen over the years, there is a simple thumbnail graph that demonstrates the percentages of those who have tried to do the part of their idea's development themselves that is best left to the experts:

Doing It Themselves (%)

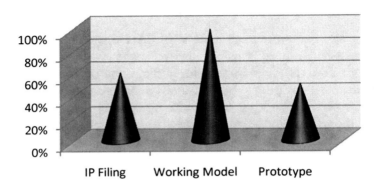

- 50% of the intellectual property protection has been written and filed by an individual or if a company, by the owner or owners

- 90% have made their own working models

- 40% tried to make final prototypes themselves

The cases that failed on every occasion to achieve any form of commercialisation were where the owners of the idea filed their own IP protection or attempted to make credible prototypes themselves or did both. The solution on both counts is DON'T! There were two reasons why they did so; first was the element of stubbornness, thinking that they knew better and never taking any advice even though in some cases, the advice was free. Secondly, was cost; they could not afford to seek expert advice. For some of the owners of these ideas, both stubbornness and incurring no cost combined to form a toxic blend of abject failure in their idea's development.

It begs the question, why pursue the idea if for whatever reason you will not employ the use of experts? Yes, you may think that you have saved money, but in fact the reverse is true, you will have spent some money and an inordinate amount of time in trying to develop an idea that is most likely doomed to failure.

Agreed for most, money is not in abundance and therefore incurring costs to develop their idea is a critical consideration. I offer one suggestion here – read Chapter Nine; "Seeking initial, early stage finance for project development" and in doing so, re-think your approach to the development of your idea. If you are however, one of those of a stubborn disposition who believe you can do it all yourself, please feel free to more than likely fail!

Using the experts - IP protection and filing

I know that it sounds like a needle stuck in the groove of an old LP, but use a patent attorney for the filing of your intellectual property protection. You do not need a confidentiality agreement in place; they are bound by a professional code of confidentiality. Revealing your idea to them does not put your idea into the public domain.

The first consultation will be free and they will give you 20-30 minutes of their time to review your idea. Whatever you do, don't over-enthuse; it is so easy to do, so be practical and objective. Prepare for this wisely using the 20-30 minutes to maximum effect, providing them with:

- An explanation as to how the idea came about and the purpose your idea serves – it's very easy to give your life story here and use up all the time allocated, so keep this brief almost in bullet point fashion and allocate no more than three minutes to this

- Any drawings

- If it is practical in terms of size or weight to do so, take your working model

- Your *thorough* IP search results – this will show that you've done your homework and to some extent, they will base their decision as to whether making a filing on your behalf is appropriate. Patent attorneys will not normally do their own IP searches unless they have a specific query relating to your idea that they could answer by spending ten minutes of their own time doing so

- The basic market searches to show where your idea sits in the market. This exercise gives them a good rounding as to the concept overall. Price point detail is not important

- Your "ripping up the rule book" train of thought; what else could your idea possibly be used for?

Yes it appears to be a lot to get through in a short time, but if you provide the patent attorney with this information, preferably in the order shown and stick to the discipline outlined, they will be armed with enough information to make a considered judgment as to whether to undertake a filing on your behalf. If you provide everything detailed above, in nine out of ten cases they will do so.

Of course each idea is different and the fees charged by a patent attorney to write and file your application will vary depending on the complexity of the idea. A typical charge for writing and filing a UK patent application would be somewhere in the range of £900 - £3,000 plus the UK IPO fees. Please note that these are not finite costs; if the idea is extremely complex requiring several drawings and a long list of claims to go into the patent, the cost could be several thousand pounds more.

The patent attorney may be able to tell you the cost at the initial consultation, but quite often, they will come back to you in a day or so with these details.

This information will provide you with the first substantive amount for your overall cost planning.

A word of warning, these costs cover a UK application; if you are going to file a patent application in other countries at a later stage because your basic market searches show a market for such, then consider the following:

Within twelve months of your UK filing date, you have to decide in which other countries if any you would wish to file an application and it has to be filed within that time. **After twelve months of the UK filing date, you lose any right to file an application for your idea in any other country.**

If you wish to do a broad-sweeping filing to cover the rest of the globe within the twelve months then a PCT (Patent Co-operation Treaty) filing is appropriate. This is a blanket-coverage of all the countries who subscribe to the Patent Co-operation Treaty, a total of 139 countries. This is not a world patent but a means by which each of those countries is notified that you have a UK filing date which is then known as a priority date. Using a patent attorney, a PCT filing will cost around £2,500 or possibly more. Once you have filed a PCT application within twelve months of your UK filing date, then you have 30 or 31 months (depending on the country) to file an application in the specific countries of your choice – this is generally called the "national phase". This is where it gets costly; costs vary from country to country and following on from your PCT filing, filing your patent for example in the USA could cost £1,500 and filing in countries such as Japan or South Korea could amount to £3,500 in each of those countries as the application has to be translated into the language of that particular country.

Please note that not all applications are filed in the UK initially, you may decide to go for a US or in Europe (EPO) patent application from the outset (depending on what your basic market assessment reveals). This invariably will cost more than an initial UK filing, so your decision may come down to cost consideration. Your patent attorney will be able to advise you on this.

Using the Experts - Design and development

Chapter One "Setting out" suggested that where you may have some difficulty in building a working model for whatever reason, then employing an expert in their field (subject to the model looking viable if you had their help at that stage) for a relatively moderate cost early on could be the way forward. As with patent attorneys, initial consultation with design experts will be free. At this proof of principle stage, then all you are looking for is input so that armed with that expertise, you will either be able to make the model yourself based on their suggestions or they will be able to make the respective components that are outside of your scope, so contributing a key element to the overall working model build.

Generally, only use them if you are confident that their help will contribute to a successful working model at a minimal, appropriate cost that you can afford.

Otherwise, I would suggest that you forget the idea and once again, move onto the next one.

Chapter Nine

Seeking early stage finance for project development

Progress so far

So far there have been various stages of your idea's development which have either followed one after the other or have been operating in parallel.

You will have:

- Made a working model with or without the help of expert help at a minimal cost

- Consulted a patent attorney and subject to their having agreed to undertake the writing of a patent and the filing of the application, agreed a cost with them to do so

- Sought consultation, possibly with a university to design and develop your idea from a working model through to a first stage prototype which will conclude with a final prototype ready for commercial exploitation. The development schedule and cost will have been mutually agreed

- Held back on investing in extensive market analysis at this stage

Apart from the working model build, you will not have incurred any costs for the patent filing or the prototype development yet; you are gathering up all the information to review the project costs for getting your idea into its advanced stage of development and for commercialisation. Although having carried out basic market analysis, extensive market analysis will follow later, once you are well into the prototype development of your idea.

It now begs the question - How much can you afford in financing your idea's development further? Maybe you can afford to finance it in its entirety or more likely, you can only afford some percentage of the overall cost to get it to the advanced stage of development? If you can afford to finance these costs yourself, then fine, but in the vast majority of ideas that I've seen, this certainly isn't the case.

This is where seeking early stage finance could be your only option to get your idea to the final prototype stage of development together with a filed, robust patent and if necessary a registered design and trade mark. You may be able to borrow the money, obtain a bank loan or a grant to finance the development. Generally, if you need finance to develop your idea, then doing so successfully at this stage is often far harder than if you have a fully developed and protected product ready to be commercialised. The main reason is you only have a belief that there is a market for your idea (your basic market searches help to endorse your belief), yet you do not have any orders or anyone interested in a licensing agreement and it has still not been developed into a final prototype where it could still fail due to technical obstacles.

For this reason, if borrowing the money, obtaining finance through a loan or receiving a grant is as far as you're concerned out of the question because you want to own 100% of the idea, then here is one of idea development's golden rules and it is up there with the most important of them,

A good percentage of something is far, far better than one hundred percent of absolute zero!

In other words to concede a share of your spoils to get your idea financed; to get it to where it needs to be is far better than giving nothing away and by trying to do it all yourself on a shoe string, possibly being doomed to failure. Believe me, there are many who take the latter option as they take the view that if someone invested £10,000 in their idea for a share of say 20% of future rewards, then that investor could make for example £100,000 over a period of time if the idea was successful. Their mind set is,

"Why should the investor get ten times their money back if they have done very little for it and I've done all the work?"

The answer is simple – they have invested faith by way of putting their hard earned cash into you and your idea; remember, they don't have to. Put it this way, they could spend that on their kids or family in enjoying life but no, they have invested it for all intents and purposes in you. Whichever way you look at it; that is a risk!

Yet the reward for the risk still breeds resentment, but why? It means if the investor has made £100,000 on a 20% shareholding, then the owner of the idea with 80% has earned £400,000! I ask here, who would be unhappy with that? They do not see what they are getting for receiving that investment of £10,000, more the fact of what "I'm giving away" and that this investor has earned ten times what they have put in.

Sadly, because of this attitude many possible good ideas have disappeared down the pan!

It is worth noting however if, once you've completed the final development of your idea and you are looking then to manufacture yourself, you may need further investment for tooling and marketing amongst other overheads to get your idea into the marketplace in its multitudes. If this was the case, you may have to concede a further percentage of shares in exchange for further funding. This is explained in Chapter Twelve; "Putting your idea into production".

Planning project costs and time line

Armed with the information that you have gathered from both the patent attorney and the specialists who will develop your prototype, it is now time to plan the project costs and the time line associated with the development of your idea. Although each case will be unique, we will produce a case study.

Boiling over with ideas

Sharon Lawrence and Jo Hall had devised a novel saucepan and base which once the base was connected to electricity, it would heat up the saucepan more rapidly than conventional hobs and use less power for the equivalent heat output. The principle was similar to the way a kettle plugs into its base. One of the benefits was that the base could accommodate three different sized saucepans, although only one at a time. It was their intention that the bases could be bought separately and the great advantage was that once the saucepans and bases had been used, they could be stored away so precluding the need for a hob. In properties where space was at a premium, this was a distinct advantage.

They had built a rough and ready working model which was extremely useful in getting both the patent attorney and the design specialist to understand the concept in its entirety.

To protect their idea and to get a final prototype, the costs and the associated timeline were mapped out. Sharon and

Jo felt from their basic searches that there was a potential market for the saucepan in the UK, Europe and the USA and the patent attorneys had quoted on the basis of filing for IP protection in these geographical regions.

The design faculty at their local university drew up a plan of action together with associated costs. They felt that to save money, a first prototype should be built for one saucepan alone. The first prototype would have a look that would appeal to potential purchasers of the product and would show up any possible issues in the design and function of the device.

Once the first prototype was finished and utilising the knowledge gained from this build, a second or in this case a final prototype would be built. This included the possibility to accommodate all three sized saucepans, using production materials where possible resulting in a final aesthetic look. This final prototype would be ideal in seeking finance to manufacture or seeking a licence agreement.

Peter Ford of De Montfort University comments,

"To be presented with a working model is extremely useful in understanding our clients' thinking; it helps to crystallise the clients' thought processes and often helps us to see more clearly what they perceive to be the end result. It is also important in mitigating some costs as they may have undertaken work themselves that we would otherwise have to incorporate in the prototype development." And goes on to make the point,

"In the case of Sharon and Jo's idea, making the first prototype based on one sized saucepan reduces the costs at this stage and helps to iron out any technical issues, either unforeseen or anticipated before embarking on a production-ready final prototype."

The costs - Patent attorney

In this case study to keep costs low at the outset, the patent attorney suggested filing a UK patent application first followed within twelve months by the filing of a US and EPO (European) patent application if it became appropriate to do so.

	Cost	When to File
UK Filing	£1,800	Initial
US Filing	£1,600	11 Months after Initial
EPO Filing	£2,700	11 Months after Initial
Total in First 12 Months	£6,100	

There would be other costs; if when the UK application is examined, the UK IPO examiner could cite any other patents that the saucepan patent application may in his or her opinion infringe, then the patent attorney would have to argue the case against the examiner's findings.

In addition, a substantive examination of the patent application would have to be requested and paid for within six months of the application being published. On the basis of these assumptions, the patent attorney suggested that a contingent sum of £600 - £1,000 should be considered. Sharon and Jo decided to err on the side of caution and put £1,000 into their cost planning.

Whilst they felt they were fairly close to mapping out all the claims for their patent, Sharon and Jo resisted the temptation to file whilst they only had a working model. They decided to wait until the point in the prototype development where they were satisfied that any issues arising had been identified and resolved. Delaying the filing up to this point in the development would also give them longer to covertly develop and exploit their idea before the patent application was published 18 months after filing.

The costs - University design faculty

The university put together a development schedule along with costs for creating a first prototype and then a final prototype.

First Prototype	Cost	Time
Scoping Out, Technical Feasibility and FMEA	£2,500	6 weeks
Build First Prototype including Material Selection, Ergonomics and Initial Aesthetic Design	£3,500	2 months
Testing of First Prototype	£500	1 month
Second / Final Prototype		
Build Second Prototype Incorporating Technical Amendments from Testing of First Prototype, Final Material Selection and Aesthetic Design	£5,000	10 weeks
Testing of Second / Final Prototype	£500	2 weeks
Total	**£12,000**	**c.7 months**

The university proposed that Sharon and Jo pay a 25% deposit on commissioning of the work with an interim payment of £4,000 upon completion of the first prototype and the balance of £5,000 paid once the work was complete at the end of the seven month project. Would the university make a government funded contribution to these costs? Read on.

The total cost of IP protection and product development amounted to £18,100 plus a contingency of £1,000 with the patent attorney; a total of £19,100.

Assuming that the UK patent application is filed half way through the development of the second prototype (five months from the start of scoping out) then the twelve month window in which to file for US and EPO patents along with requesting a preliminary examination begins then. This total spend of £19,100 is over approximately 17 months. Sharon and Jo plotted a simple cash flow forecast over those 17 months, working out what money would have to go out in which month. A point to bear in mind is the costs quoted will be exclusive of VAT which will have to be added to the above. If you are VAT registered, then of course that can be reclaimed and even if you are not VAT registered, there is the possibility that you could reclaim the VAT retrospectively if you eventually register for a VAT number, although it is time limited as to how far back you can reclaim once registered. Always consult your VAT office prior to reclaiming in such circumstances.

Before Sharon and Jo committed to the prototypes, they examined their finances and between the two of them, they could rustle up savings of £5,000 to put towards the project cost of £19,100, but where would the rest of the money come from?

Whichever form of funding you seek, you will have to fully disclose your idea and its history of development along with your future aspirations – if you don't you won't get funding; it's as simple as that. Unless you have a confidentiality agreement in place with whomever you discuss funding requirements, you render yourself completely exposed and your idea is then in the public domain. ALWAYS ensure such an agreement is in place and signed prior to any revelations of your idea.

Funding the project yourself

Maintaining 100% ownership of your idea and reaping the subsequent rewards if it is successful by funding the full costs yourself is often the preferred route. Yet for most, it is financially impractical or more often than not, they are not prepared to put their money where their mouth is and risk their own cash! It begs the question; if you have set a plan aside for the development and possible exploitation of your idea and there is passionate belief in its success, why not do so?

Human nature or pressure from family or business colleagues says "no, don't risk it", yet if you turn to others for funding, they will ask you why you are not prepared to put in the money yourself when you are sure that your idea will work and be successful. If you cannot or are not prepared to fund this development yourself, you have to be very careful about your answer when they ask that question; many who seek funding trip up on this question and are left empty handed. This issue will be returned to further on, but first what if you are prepared to fund your idea's development and IP protection yourself?

Maintaining 100% ownership of the idea
Looking at the case study of Sharon Lawrence and Jo Hall, what are their options if they wish to get some funding together one way or another yet maintain 100% ownership of their idea? They are the sort of people who will consider every option before making a decision and their first option is getting the money together themselves.

To maintain 100% ownership, one of only two ways to do so if your own funds are insufficient to cover the cost of the development is by way of secured or unsecured loans that are repayable. This way, decisions to fund you are often reached far quicker and require less detailed input than other sources of funding. The other way of maintaining 100% ownership is grant funding which will be addressed in the next section.

You should consider loans from these possible sources:

- Friends and family – probably the quickest of all funding solutions. There are many instances where an idea has been funded by such. Your friends and family are probably on board already in terms of supporting you in the development of your idea so they could well be prepared to loan you the money to see the idea complete its development cycle. It is important though to remember an old saying,

  "Make friends of your enemies, but don't make enemies of your friends!"

  It is fine whilst everything is going along smoothly, but if for whatever reason, the idea fails in the development stage due to unforeseen issues, you are left with a failed project and indebtedness possibly in the thousands of pounds to those who have supported you; you could be treading on glass if this is the case.

- Bank or finance house loan – possibly the "cleanest" way to go about financing your idea's development; you have no other individual involved apart from yourself and the lending institution. The downside of course is that you will have to start making regular repayments once the loan is in place. If you are considering this route, make sure to put these repayments into your cash flow planning during the development cycle and beyond.

Ideally you would want the loan over as short a period as possible, but the shorter the period, the higher the repayment instalments. A popular way to finance the development costs is by having a loan secured against your mortgage, assuming that you have enough equity in your property against which to borrow. This is normally spread over quite a number of years and whilst it is probably the most expensive sort of loan in the long term, the monthly repayments are far lower than shorter term loans, so easing your cash flow burden.

Banks and finance houses look at providing funding based purely on you or your company's ability to repay the loan taking into consideration credit history and worthiness, the viability of the idea along with its commercial potential, all of which must be detailed in a robust business plan. They will not necessarily understand the intricacies of the idea or all its commercial attributes. In a way they are like patent attorneys who work to a set of rules and will in general not extend their knowledge or risk beyond them.

In committing to any form of loan from a bank or finance house, that old saying rears it head; "read the small print!" and know what you are committing yourself to. They say that in the world of gambling, you should only bet on something if you can afford to lose the money. Well I would say look at your idea as a "bet" whether it appears to you to be an odds on favourite, evens or even higher odds and ask yourself if it loses (fails) could you afford the punt (loan).

Maybe the answer is no, but if you consider it to be an odds on favourite to win, then you're prepared to take the risk. To many, that is considered foolhardy, to others, that is considered entrepreneurship – only you can decide which hat you wear!

- Small Firms Loan Guarantee Scheme (SFLGS) – This is a loan scheme available from banks to those who do not have any assets against which to secure a loan and a large percentage of the loan is guaranteed by the government to the bank. You have to have a viable business plan for the loan to be granted. As you require finance to develop your idea, it is highly unlikely that you will have a fully developed business plan demonstrating sales and income, so it is most probable that you will not be in a position to apply for such at this stage but may be able to do so later if you are going to make and sell your fully developed idea. The SFLGS will be explained in more detail in Chapter Twelve; "Putting your idea into production".

Funding assistance from universities, government and RDAs

Having looked at the loan opportunities, Sharon and Jo decided to weigh up the other options. They were reluctant to give away a percentage of their idea to fund its development, but had not closed the door on that option. With a little research, they mapped out further opportunities that may be available for funding their idea's development.

The first ones that they looked at were government and Regional Development Agency (RDA) funding initiatives.

Government backed university funding initiatives
Across the UK there are millions of pounds of government money that back universities in the development of individuals or companies ideas. The funding is normally a mix of what is called capital funding, to provide the universities with state of the art equipment and the other is revenue funding; funding which can be spent on clients' projects. It is an excellent way of getting your idea funded through its development stage, but before you start jumping up and down with excitement thinking that it's all paid for, think again! There are a set of European laws which govern such funding called "State Aid". In its simplest form, State Aid rules mean that the development cost of any project can only be funded up to a maximum of 50% of the project cost. There are a lot of criteria attached to this funding, some of which are complicated. I have spent endless hours in the past trying to ascertain from those who should know in RDAs and central government in London as to what the finite interpretation is of these rules and no one can give a straight answer that is consistent with the answers of others. So if you get confused along the way, you have my sympathies.

In the case of Sharon and Jo's project with a cost in the design development stage of £12,000, they could be eligible for up to £6,000 of grant funding through the available initiatives. Please note that funding is not automatically set at 50%; that is the maximum and is at the discretion of the university involved. De Montfort University's Improving Business by Design and Resource Efficient Design (RED) are two such design and funding initiatives.

When approaching a university with your idea, they will normally volunteer the information on the funding initiatives available as once they have been granted the funding, in its simplest of terms, they have to spend it! In spending it they have to hit the targets (outputs) that the funding body set when granting the money, so in nearly all cases, they are willing to assist you wherever they can. If the information is not forthcoming, which is unusual, always ask them what is available. In some cases, specific faculties within universities may not have funding available from time to time due to not meeting specific criteria for a set of funding regulations or that a funding initiative that they are running is coming to a close.

Government Grant for Research and Development
This particular form of grant varies in the amount of money available, the type of development required and as to how long the development of your idea lasts. It is available to individuals and small and medium-sized businesses in England to research and develop technologically innovative products and processes. There are similar types of grants available in Scotland, Wales and Northern Ireland, but with slightly different criteria applying.

There are four types of grant available, each supporting different types of research and development projects, and each requires the applicant to make their own contribution to the project costs. The three most appropriate are shown below:

- *Micro projects*

  These are simple projects with low-cost development that last no longer than 12 months. So long as your business employs fewer than ten people, a grant of up to £20,000 is available.

- *Research projects*

  If your idea requires an investigation into the technical and commercial feasibility of innovative technology and would take between six and 18 months to do so, then a grant of up to £100,000 is available. Your business has to employ fewer than 50 employees.

- *Development projects*

  If your research and development has been completed and you now wish to develop a pre-production prototype of a new product that involves a significant technological advance and to do so, will last no less than six months and no longer than 36 months, then a grant of up to £250,000 is available. You have to employ less than 250 people.

Further information on these grants is available from your local Regional Development Agency; there are nine in England. The nearest RDA to you can be found at:

http://www.englandsrdas.com/home

Outside of England, information can be obtained from the websites of the Scottish Government, the Welsh Assembly Government and Invest Northern Ireland.

It is important to note that with all grant funding, there is a requirement for you to match the funding with your own money to a certain level; you won't get 100%. Money that you have already spent on developing your idea will not be counted as part of the match funding; it is only future project costs that are considered.

Another point is that you cannot go to different organisations to part fund your idea through grants so that the sum total funded by the different bodies amounts to a 100% of the cost of your idea's development, although I have seen attempts at this on several occasions. This breaches State Aid rules which were mentioned earlier. Most organisations when funding you and your idea will insist that you sign a State Aid declaration form stating how much in grants you have received for all projects for which you have sought funding over a given period of time and there is an upper limit. Exceeding this limit can mean you could possibly have to pay back the funding that you have been given.

Funding from private sources and other parties

This can be quite a difficult form of funding to obtain, particularly in the development phase of your idea; but if funding is forthcoming, it can be the most beneficial way in which to move your idea forward. Make no mistake, if you seek funding from private sources, you will have to give away a percentage of your idea. In doing so, you are not borrowing the money; it is being given to you in exchange for a share in the future rewards generated by your idea. This is called "equity funding".

At this development stage, only one source of private funding will be looked at, although there are others which will be explained in Chapter Twelve; "Putting your idea into production".

Business Angels
Business Angels re individuals who could be someone you know personally or via a contact. They could also be a member of a business angel consortium. A business angel consortium is made up of individuals and/or syndicates. A syndicate for all intents and purposes is made up of individuals who collectively invest in an idea.

They are normally only interested in investing what is to them small sums of money, but to you could be a lifeline. On smaller projects, it will normally be an individual who will invest and very rarely will they invest less than £10,000, but will invest larger sums if the idea seems viable.

Syndicates tend to invest in larger sums and could consist of three, five or any number of individuals. If for example, your idea required £125,000 investment to develop it to its full commercial potential, then this money could come from syndicated funding within a business angels consortium, comprising say five individuals investing £25,000 each.

Sums of £10,000 up to £100,000 are quite common for each of them to invest in what they perceive to be good opportunities.

As this form of funding is private, it is not affected by State Aid rules and for this purpose is seen as your own money as a contribution towards the costs of development.

What do you give away and what do you gain?
In general, you can look to be giving away a percentage that may be in the region of 25%. This is by no means an absolute figure at all, as each idea that is to be developed will vary in terms of level of investment, technical difficulty and size, time to develop, market potential and immediacy (how quickly it can be got to market). There can be a complicated set of rules and agreements behind their investing in your idea which you have to look at carefully with a professional body such as an accountant. Some of the typical subject matter to look out for in such an agreement is:

- The sum of money that they will invest

- When they will pay such a sum (in one payment or in instalments across the development cycle)
- The percentage of the idea that they will own
- The percentage of the idea that you will own
- When they have an option to exit
- How and when they will be rewarded for their investment (their reward often taking first preference over any future reward you may get)
- Their right to have first refusal should you require further investment at the commercialisation stage
- A full indemnity for the investor so that they are protected against any liabilities arising whatsoever in the development and commercialisation of the idea
- Your right to first refusal to purchase back their percentage when they exit

These are the basics of such agreements and there will be other inclusions for sure, but it will give you an idea what to look out for.

By using an investor, you have two distinct advantages, one being that you do not have a loan that has to be paid back either monthly or quarterly over several years and secondly, they will normally be versed in the technology of your idea; they will want it to work, so they will give free expert advice and encouragement to give it the best possibility of succeeding; they will want to work with you. This expert advice, if you were to pay for it on the free market could cost £500 - £2,000 per day. Look at it constructively, this is a valuable commodity and is not just a case of giving a percentage of your idea away, but an investment on your side in a wealth of potential expert knowledge.

An investor once having invested will at some future date look to sell their interest in the idea (normally, you would have first option to purchase back their interest). This selling of their interest is called an "exit"; it normally occurs within three to five years of investing and they hope to do so with a healthy profit. Of course some lose on their investments, but they would normally use an element of this loss to offset their future tax liabilities.

Sharon and Jo had decided on their funding for their idea's development. The university offered to contribute via one of their government funded projects and with Sharon and Jo's contribution, there remained a balance which was funded by a business angel. He happened to be a friend of Jo's uncle who owned a chain of kitchen shops; this was an easier and more informal route than going through a business angel consortium and they gained his expertise in that particular market. They had considered a loan for the balance, but they felt equity funding was by far their best option.

Planned Budget	£19,100
Sharon and Jo's Funds	£5,000
University funded contribution	£4,000
Business Angel (for 30% shareholding)	£10,100
Total Funding Achieved	£19,100

It's interesting to note that the business angel has funded over half the cost of the development but for less than a third of the equity in the idea, giving Sharon and Jo 35% each; they still maintain control of the idea with 70% between them.

What gave Jo's uncle's friend the confidence to invest? Quite simply, he was impressed by:

- Sharon and Jo's determination and commitment
- Their organisation and document trail
- A commitment by them to spending £5,000 of their own money
- The willingness of the university to part-fund the project

At this stage in your idea's development, you may find the funding checklist a useful aide memoire.

Funding Checklist

Project Costs	Amount
Patent Attorney	£
Design Institution	£
Legal / Professional	£
Other	£
Total Costs	£

Funding Channels	Amount
Yourself	£
Friends / Colleagues	£
Business Angels	£
Bank / Finance House Loan (inc. SFLGS)	£
Government Backed University Funding	£
Grant for Research and Development	£
Other	£
Total Funding	£

Chapter Ten

Drawing development and protection together

By now, hopefully you are firing on all cylinders; things are really underway. Funding is now available to commit to the final development phase for your idea, enabling you to draw prototype development and protection together. Before doing so, as there has been a period of time since you did the original IP searches, dig out those previous search results from the IP databases and using your original keywords along with any new ones that you can now think of, do the searches again. As emphasised previously, do not skimp on the time and effort you put into this; do it thoroughly.

If for example, it has been three months since you undertook your previous searches, then there are three months worth of newly published patents and registered designs that were not previously in the public domain when you last searched. If there are no new results that merit either concern or consideration; that's excellent, but if there are areas for concern, call a quick meeting with your patent attorney to discuss how to circumvent those concerns.

As your financial commitment will shortly be running into the thousands of pounds, it may well be worth using the Research Service team at the British Library at research@bl.uk. For a sum of around £150, they will be able to undertake a patent search on your behalf and they will need a description of your idea along with a list of your key words. They have access to various databases and normally within a week, will produce a report containing enhanced titles or summaries of the patents that may be in a similar field to yours. The team are extremely helpful but will not furnish an opinion as to whether infringement is a possibility or not; that is for you and your patent attorney to discuss.

Further information on the intellectual property research service offered by the British Library can be found at:

http://www.bl.uk/reshelp/atyourdesk/docsupply/productsser vices/researchservice/how/index.html

Prototype Development – the first meeting

The patent attorney can be parked on the side for the time being; the next meeting will be with the design house or institution that you have appointed to undertake the prototype development work.

Just to remind you, I refer all the way back to the beginning of the book,

"Be invitational, not confrontational"

This will be the first time in your idea's development that you have let go of total control and management of your idea. Once the prototype development is underway, there will be a temptation to phone up every few days to see if those who you have commissioned are getting on with developing your idea and if there is any spectacular news; after all "you are paying them now, so they should be getting on with it!" I've seen irritation grow rapidly with people who have "not heard anything for a week now!" and to my angst, have been on the receiving end in a few cases and have to say that it drove me absolutely potty and creates a negative collaborative effect all around.

There are three points that I'd like to mention here regarding those observations,

- Look at the time agreed to develop your idea; in the previous chapter, Sharon and Jo's idea was planned to take seven months. Prototype development is not a quick fix

- If using a university to develop your idea, the people that you are dealing with have various other commitments such as teaching students and producing academic papers (that's why prototype development time is planned to incorporate such)

- Whoever you have appointed to develop your idea, yours is not the only idea that they are developing; it is one of several or probably many

This is not meant to be condescending, but I am a great believer in dialogue only when there is something to say or a decision has to be made. I would suggest that at the first meeting, it is agreed as to when during the prototype development contact is made or meetings convened. There is absolutely no harm in you contacting those who are building your prototype a few days before each agreed date of contact or meeting, just so that it gives them a gentle nudge and reminds them that you are on the ball in a constructive manner.

That first meeting will be to scope out what exactly needs to be done in the development cycle, looking at the technical feasibility and possible FMEA implications. Take along all the documentation that you presented at the initial free consultation together with the quote and development schedule that you have received. Take a sheet of bullet points also that list how you understand the development to be undertaken and as to what the final result will be. This is so important; so often, there is ambiguity in the interpretation by both parties that if not removed at the beginning, can cause all sorts of hassles and disagreements later. Of course, the finite outcome is not known, as by its very nature, it is development and your idea will change, but if this clarity is included in the mutual understanding from the outset, then there is every chance that progress will be smooth.

It would help wherever possible if whoever you appoint has all the resources in-house to develop your idea through to the final prototype. This is not always the case and they will employ external resources where needed, the slight disadvantage is that they lose some element of control, but it is their responsibility to ensure they maintain such. Once again, if this external resource is being employed, make sure that they are covered by the confidentiality agreement that was drawn up between yourself and those who are developing the prototype; you can't afford any loopholes at this stage now that serious money is involved.

Consultation between design experts and patent attorneys

Out of courtesy, inform the patent attorney that you have all the funding in place, that you have commenced the prototype build and that you will contact them when the patent application (and possibly registered design) is ready to be drafted. This will keep them keen and they will be more than willing to help wherever they can, knowing that you have those funds. I have experience of patent attorneys spending an inordinate amount of time chasing invoices where they have done the work and the client, because their idea has not worked, have been reluctant to pay. Yours is the type of work that they want, not the latter.

When it eventually comes to the drafting of the IP protection, it is a good idea to get the patent attorney and those who are building the prototype together for a meeting at some point, even if only for half an hour. There may be little nuances within the prototype i.e. the way in which one element mechanically functions with another or indeed use of materials which may not appear particularly relevant to you, but for want of a better word, once explained in "techno speak" could affect the way in which your patent is written for the better.

There may be a requirement once again for the two to meet once the first draft of IP protection is written to ensure that the patent attorney fully understands the explanations previously given by those that are developing the prototype. Another point is there could have been subsequent developments in the prototype that merit inclusion in a redrafted form of the IP protection.

These meetings may cost you a little time and money, but history shows that they are extremely worthwhile in writing a solid, robust patent.

Your idea is much more than just an idea now, it's rapidly becoming a product.

Chapter Eleven

The big picture – assessing the true market potential

Looking back to your early market assessment

From the outset, undertaking basic searches to see whether there is any other idea like yours whether on the intellectual property databases or through simple market analysis was a cornerstone in your decision making; "have I got an idea with potential?" or "shall I forget it and move onto the next idea?"

Either way, your costs at that stage were zero or at worst a small sum and reading this chapter of the book means that you will have made the decision that your idea has that potential; after all, your idea's development is funded and the prototype is close to being finalised.

In the early market assessment you will have looked at the different ways of searching the market and have used some or all of the following,

- Internet/Websites
- General Magazines
- Specialist Trade Publications
- Retail Stores
- Specialist Outlets
- Exhibitions and Trade Shows

Before looking deeper into the true market potential, revisit the information that you originally gathered together to refresh your mind as to what ideas were out there along with the sort of prices in the market that similar or comparable products were selling for.

Having done that, retrace your steps doing your basic market search once again. The reason for doing this is simple; it may now be six months, a year or even longer since you undertook those initial markets searches. In that time the wheel of new product launches will have been turning rapidly. This is useful in considering whether there,

- Needs to be any tweaks to the prototype

- Is any new information that may affect the detail in your patent before it is finally filed

Hopefully on both counts, it will not be the case, but if it is, consult your patent attorney and those who are developing the prototype to incorporate the information and tweaks.

It could be argued here that it would be better to undergo this process earlier as with the IP searches, but there is no right answer here. The internet is a rapidly moving feast and products disappear as quickly as they appear. The IP databases have a natural ring-fence around them in that information only appears for intellectual property that has been filed and there are a set of rules regarding publication and dates related to such. It is infinitely more controlled than the general internet and nowhere near as random as the internet's ever-changing content; a second due diligence IP search therefore is generally more appropriate and significant at the beginning of the prototype development, followed by another one at the final prototype development stage.

An internet market search at this juncture is to familiarise yourself with that which you did before, see if there are any significant products that should be noted and is an essential warm-up for the full market analysis.

Assessing the true price point

In Chapter Three, we looked at the three legged stool where the seat of the stool is your idea with the three legs composed of IP protection, price point and market. Each leg in its own right has to be strong otherwise the stool topples as well as your idea along with it. The IP protection leg has been found to be sturdy, but it is now time to ensure that the price point and the market legs are robust and just as sturdy. In ensuring so, you will be equipped with a powerful market analysis that will shape the future of your rapidly developing product along with its commercial worth and how it will find its way into the market.

You will have completed analysis of the price point in your basic market analysis, but there is a need to firm this up to give you factual and evidential data that will be used as,

- Part of a business plan if you wish to manufacture and sell your product yourself

- Part of a presentation to a potential licensee

There are two key questions to this price point,

- What price can the product be manufactured for?

- What price will the end user pay for such a product?

What price can the product be manufactured for and what will an end user pay?
An important aspect to consider in establishing this price is; does your product require any packaging either to deliver it, protect it and/or to display it? This of course depends on what your product is, but generally all three have to be considered. If for example, your product is a new sort of

toothbrush you will probably have to consider all of the following,

- A blister pack comprising a clear plastic front to be able to see the product on display

- Cardboard backing that on the front will describe the product in an appealing way and on the reverse will display instructions for use and contact details amongst other requirements

- A tray for display purposes that will hold say a dozen toothbrush packs

- An "inner" box – this will contain a number of trays, for example twenty

- An "outer" box – this will contain a number of inner boxes

An outer box is used to transport the product from factory to distributor or retailer and the inner box is used to keep a supply of the product in their storerooms. Each of the elements of the packaging are critical in size; a retailer will only have a certain size of shelf space in which to display the product, so the packaging of ,in this case a toothbrush, has to cater for this. Conversely, upstream of this, the outer box has to be of a size that will not overhang the palettes on which it is transported, yet will be of a size that will permit the maximum number of boxes to be transported per palette, so making transport as economical as possible. Certain retailers will require packaging that protects the product, particularly if an element or all of it is fragile. The packaging will have to stand the test of being dropped from a designated height onto the floor without any damage to the product inside.

This packaging, if your product requires it, has to be factored into the price point. There are many packaging companies on the internet who you can contact to obtain prices for packaging your product. Make sure that they are fully conversant with all the elements of the packaging detailed here, otherwise it could work out costly if one aspect of the packaging does not meet the prerequisites of transport, storage and display

These packaging costs have to be added to the cost of manufacturing the product.

In assessing the manufacturing price point for your product, the design house or university that you are using to develop your final prototype can help. They will have many manufacturing contacts and they will be able to point you in the right direction in terms of getting a quote to manufacture your product. At this stage, you are only obtaining a quote from them, not appointing them to manufacture. Get at least three quotes.

Drawing the prices together for the product and the packaging form the basis of the price point, but there are some points that have to be taken into consideration. Firstly there will be manufacturing "set up costs" for both the product and the packaging; this will include any special tools that have to be made to manufacture the product and packaging along with set up time for machines to make them. These costs are split into two; one being an initial one-off cost regardless of how many production runs are made and the other being a recurring cost every time a machine is set up and allocated to a production run.

With this in mind, these costs per unit of the product manufactured become less the more units of the product that are manufactured in any one production run as those set up costs are spread across a greater volume. You will need to obtain a "price matrix" from the manufacturer to be able to assess your ideal price point. The matrix will detail the cost per unit manufactured based on different volumes of a production run.

This matrix, coupled with a price matrix from the packaging manufacturer, will now give you a range of price points called the factory gate price, in other words the price that your product leaves the factory gate to be delivered to a distributor or end user; but which price is the best one to use? There is no easy answer, but the following is important to consider when arriving at what you consider the most appropriate price for your product,

- What is the annual market potential for your product; how many units do you realistically believe it is possible to sell? This will become clearer from further market analysis

- As a rule of thumb, if a distributor and retailer are to be used, a product sells to the end user for four times the factory gate price. This price to the end user takes into consideration the gross profit margin of both the distributor and the retailer along with VAT charged to the end user

- If you are to sell your product yourself without the intervention of a distributor or retailer, then reckon on a price to the end user of around double the factory gate price

If you are to pursue a licensing agreement, then the price matrix or matrices will be an important tool when it comes to presenting to a licensee. (See Chapter Thirteen; "Licensing, assigning and evaluation")

End users are very savvy indeed and regardless of the type of product, they will do their price comparisons, so it is important to choose a production level within your price matrices that when multiplied up to the end user price, remains competitive. A word of warning here, it is very easy to use prices from the matrices that mean your product is competitive, but does that make the volume of units of your product manufactured to meet that price an unrealistic amount to sell? If your product is a "must have need" then the end user will pay a little more, meaning that the number of units manufactured on a given run can be less. There is a fine balancing line here between getting it right and failure. Remember, you can always reduce the price of your product if desired once in the market, but it is very difficult to increase the price of a product once it is there for all to see; the exception being if your are offering a launch price or special offer for a limited period to stimulate early sales.

The price point can be dramatically affected depending on where the product is manufactured; lowest prices can possibly be obtained by having your product manufactured in Asia, the Far East and Eastern Europe to name but a few regions. Conversely, you could get it manufactured in the UK, but possibly at a higher price. There is a trade-off here; using a manufacturer in the UK means you have more control, you can meet them personally and in terms of geographical distance and cost, it is much cheaper to do so.
If your product is of a very specialist nature, then the price point may not be as critical as other products in a mass market. For this reason, a UK manufacturer, particularly a manufacturer who is a specialist in their field would probably be the best option. If your product has mass market potential and is price sensitive in terms of potential sales volume, then an overseas manufacturer may be the only option to get your product manufactured at a competitive price point, the disadvantage is that you do not

have that "local relationship" that is possible with a UK manufacturer. Another disadvantage is in some countries, manufacturing legislation is not as strict as it is here in the UK, so materials, tolerances and product durability may not be of the required standard dictated by both EU regulations and what the market desires. If you are considering getting your product manufactured overseas it would be very useful to contact:

- The embassy or consulate of the relevant country

- The British Chambers of Commerce (BCC) http://www.britishchambers.org.uk/

- UK Trade and Investment OMIS service https://www.uktradeinvest.gov.uk

- Business Link http://www.businesslink.gov.uk/bdotg/action/layer?to picId=1079717544&r.s=tl

There are various bi-lateral associations between the UK and specific countries who can help in seeking a manufacturer overseas, the China-Britain Business Council http://www.cbbc.org/ being one such organisation.

What to research and analyse

The fundamentals of market analysis for your product are to establish:

- Where the geographical market lies
- The potential size of the market
- The competitors
- Market trends

- Companies likely to be interested in entering into a licensing agreement
- Financial status of such companies
- Legislation relevant to the market

You will use a combination of these fundamentals to put forward a viable case for setting up your own business, incorporating your product into production in your existing company or licensing your product to another party.

Where the geographical markets lies
This varies considerably depending on the product and can be affected by differing economies, cultural attitudes, climate, language, legislation and exchange rates. Probably the easiest geographical market to assess is the UK – after all, you came up with an original idea which in most cases would be addressing a market or solving a problem that you encountered in the UK.

What is the potential size of the market?
There are several variables here that will influence the potential size of the market:

- What is the total number of equivalent products to yours sold per annum?

- If there is no equivalent product to yours, what is the total number of products sold per annum in the same market sector?

- How many brands are there for the equivalent products?

- What is the available market size as opposed to the total market size? Two thirds of the market may be dominated by a handful of major manufacturers with the remaining third made up from many smaller companies selling their products. It would be this third that is the available market; that which is easier to break into than the market dominated by the major manufacturers.

Who are the competitors?
If you are going to manufacture and/or sell your product yourself, which companies are going to be your competitors, how dominant are they and what is their future positioning in the market? How could they affect your market when you launch your product?

Market trends
What are the market trends for your product or your market sector over the next year, two years or even five years? Is it a stable market, a rapidly declining one or a market with great growth potential?

Companies likely to be interested in entering into a licensing agreement
Which companies are they? Do they already have a presence in the market sector in which your product resides? Are they in a completely different market, but your product could open up an entirely new market for them? Does your product complement or compete with their existing product range? Are they a company with a national market or are they truly international in their market penetration?

Financial status of such companies
They may seem an ideal potential licensee, but look under the surface; is their balance sheet strong, how does their profit and loss account look, do they appear close to insolvency or are they healthy and strong with the potential to invest a considerable sum in your idea?

Legislation relevant to the market
Are there any laws or legislation that have recently come into force or will do in the near future that could considerably influence the size of the potential market for your product either adversely or beneficially? Could this legislation be a key driver in making your product a "must have need"?

Marketing analysis databases and reports

There is an absolute plethora of marketing analysis companies – one look at the internet will prove this as such, but to get valuable, worthwhile data can be expensive, often extending into the thousands of pounds. This is a cost that you may have to incur – the reasons being,

- If looking to manufacture and sell your product yourself, this market analysis data will prove invaluable in demonstrating the possible market potential. To be able to quote established and reliable market analysis companies' data and their name adds credibility to your business plan and your proposal. At this stage, just quoting internet statistics is not sufficient – that was fine for basic market analysis at the early stage, but not anymore.

- If opting for the licensing route, when you first do a presentation to a licensee, quoting the statistics and their source adds a lot of kudos to your presentation. Remember if you are at this stage, if you rehearse your presentation well, you will more than likely be quoting market data and statistics that are beyond the knowledge of the person to whom you are presenting, surprising as it may seem. This gives you a definite advantage when doing that first presentation – you come across as professional and indeed knowledgeable about the market and its value to the potential licensee.

As previously mentioned, there is an alternative to expensive marketing databases, although it will cost you fuel in your car or a train or tube trip; the British Library Business and IP Centre. The Centre has a large array of marketing and business databases that are free to use. Databases such as FAME, COBRA and MINTEL are amongst nearly forty databases readily accessible. Some of the reports that you can obtain for free would cost nearly £3,000 if you purchased them direct from a marketing company, so it is well worth travelling down to the British Library at St Pancras, London to do your analysis. Don't worry if you haven't got the faintest idea how to use such databases, the British Library staff in the B&IPC are both helpful and experienced. Plan out exactly what you are looking for before visiting, although you will find that once there, there is probably a lot more information and data that you can access and add to your marketing analysis than you first thought.

To access the information you will need to gain a British Library Reader Pass. It is free and anyone is welcome to do this, however you do need to take two forms of ID (proof of signature and proof of home address from within the last three months) with you to the Library. Have a look on their website for details on this prior to your visit.

One thing is for sure, you shouldn't come away with insufficient information and armed with this data, will be able to put a good market analysis forward whether you are going to make and sell your product yourself or opt for the licensing route.

Chapter Twelve

Putting your idea into production

With your prototype complete, IP protection (probably a patent application) recently filed and your market analysis to hand, it is decision time, if you have not already made one, as to how you are going to commercialise your product. Simply put, as mentioned before, there are three options,

- Licence or assign your product to a company

- Set up a new company to manufacture and sell your product

- Incorporate your product into an existing company if you already have one

Each has their own risks and benefits.

Licensing or assigning your product

Successfully licensing your product to a company means in its simplest terms that you sit at home waiting for the royalty cheques to drop through your front door several times a year. Or you could assign your IP to a company for an up-front one-off payment and you can now disappear into the mist, with a very healthy bank balance. In both circumstances, how wonderful! These routes to commercialisation are the ones of lowest risk; after all you have now done all the hard work, having invested your time and money and now it is pay back time! Yes it is true, a successfully concluded licence agreement or assignment (although complicated to conclude) is one of the most satisfying ways of earning money, but there are downsides,

- You will not make as much money per unit sold as you would if you made and sold it yourself

- You no longer have total control of your product and the associated IP

- Your licensee may fail miserably in its obligations to perform under the terms of the licence?

- The company to whom you licensed your IP might go out of business

Licence agreements along with assignments are covered in Chapter Thirteen; "Licensing, assigning and evaluation".

Incorporating your product into a new or existing company

Such questions may persuade you to make and sell your product yourself, particularly if you are well versed in how to both make and sell it. If you have made this decision it will either be within your existing company, or if you do not have one, within a new company that you will set up. As with licensing, there are downsides,

- It will take many long hours a week in commitment, often seven days a week

- You may have to employ staff with certain specialisms to make the product a success. Will you get on with them and will they have the same vision and commitment as you?

- You will almost certainly need funding to succeed; can you afford to pay back a loan?

- If you are equity funded, have you got a fair deal, or have you sold your soul to get the money?

- What if, after say a year or two years, it all goes "pear shaped" and you go out of business with personal guarantees now being called upon?

I know that all sounds doom and gloom, but these points have to be considered when deciding whether to take the step and "do it yourself". This chapter looks at exactly that; "doing it yourself".

The upside of doing it yourself is that you are entirely or mostly in control of your product and your destiny; another plus is that you will make more money per unit sold than through a licensing agreement; you have more freedom to develop the product and the business further and in the direction that you choose.

We will look at the setting up of a new business to make and sell your product. Many of the issues explained will apply to an existing company incorporating a new product into its business.

Where to start

With filed patent application (possibly registered design and trade mark applications as well) in one hand, prototype in the other and market analysis sandwiched somewhere in between, what do you do now? You don't want to licence or assign your product and its IP as you are confident that you can make a better job of it than anyone else could, making more money along the way. This is where the trusty business plan comes into its own; make no mistake, you will definitely need one, even if you are fortunate enough not to need funding. This chapter does assume however, that funding will be an essential element in you making and selling your product.

Self-employed, partnership or limited company?
This is a question about which you will definitely need to see an accountant. It is such an individual issue depending on you, your partners and your product. An accountant will be able to advise on all the implications of each, but some very basic pointers are given below and there are far more legal and financial implications to each than indicated here.

Very broadly speaking, a self employed person is someone whose gross income is based on sales, less the cost of those sales and any associated overheads. From this a net income is calculated after deduction of tax and national insurance. The benefit of being self employed is if your gross income is high, then you can possibly earn a higher overall net income than if you are employed by your limited liability company. One of the risks of being self employed is if any liability occurs that results in legal proceedings against your business, it is against you as the individual and does not offer any protection as can be the case with a limited liability company. Paying tax and national insurance contributions occurs twice a year and so there is a cash flow benefit (so long as you haven't spent it!).

In a partnership, exactly the same applies as above, the difference being that the gross income is obviously split between two or more individuals depending on how many are in the partnership. Any liabilities could be shared as well.

Setting up a limited liability company with you as a director means that your net income is paid after deduction of Pay As You Earn (PAYE) which is paid to the Revenue on a monthly basis. You would normally be a director if it is your own business and a director is deemed to be an employee. If the business is going well and earning money, you can pay yourself a bonus which again is subject to PAYE. If it is your company, you will have a percentage of the shares in the company (maybe a high percentage) and again if the company performs well, you could pay yourself a dividend as a shareholder. If the company makes a profit, that profit is subject to corporation tax.

There can be a lot of tax implications with a limited company depending on how well the company performs. As the company is a limited liability company, it in effect means that there are limitations to any liabilities, financial, legal or otherwise to which you might be exposed. It is a complicated area because as a director, you have certain director's responsibilities; if for example your company was sued by another party, it is usually the company that suffers from the outcome if not favourable, but if you as a director were found negligent, you could be financially and legally accountable as well.

As mentioned, these are very rough guidelines and it is for the reasons above that you should not decide which way to go without consulting an accountant.

Preparing a business plan

An important rule in preparing a business plan is to ensure that you have good robust information and data, market statistics, financial information, equipment needs along with premises requirements and probably most importantly, a provisional order book for your product. There will be a strong temptation to shoehorn all of this data and information into a business plan that "looks good" and suits you – DON'T. Be realistic and don't mix optimism with delusion; the questions that you will be asked, particularly when seeking funding will see right through this approach. If you do take this approach, it will be very hard for you to "hold it together" when presenting to potential funding providers and you could well forfeit any chance of successfully obtaining the funding needed.

You will work several models of the business plan before you finally settle on the one that is realistically the most achievable. I don't wish this to sound alarming, but once you have finalised the business plan, you will have to know just about every page of it in its entirety off by heart; if you don't know your business or potential business inside out, why should anyone trust you with their tens or hundreds of thousands of pounds?

The business plan

A good business plan should include the following:

- An Executive Summary
- Background
- Product(s)
- Markets and Marketing
- Manufacturing
- Key Personnel (Management Team)
- Funding Requirements
- Schedule of Overheads
- Schedule of Income
- Three Year Cash-flow Forecast
- Three Year Profit and Loss Forecast
- Three Year Balance Sheet Forecast
- Appendices

Each funding body will have their own nuances in the way that they wish a business plan to be presented, so the list and order above may sometimes vary.

An Executive Summary

This is probably the most important page in the whole of the business plan; it is the page that potential funding bodies will read first and more often than not, it is the only page. Why? If this page is not well written, they will read no further. This is your first and possibly only "bite at the cherry". An executive summary is your sales window and it HAS to be sharp, factual and written in a way that invites the reader to turn to the next page in the business plan. There is a possibility that ten different people will read it in a different way, so to reduce this, get this page spot-on without ambiguities. An executive summary should rarely ever be more than one A4 page; these funding bodies may be looking at dozens of these proposals every day, so reading two pages does deter them.

The Executive Summary should include the most relevant and important extracts from the rest of the business plan,

- When you expect to turn in a profit and anticipated profit after three years
- Background
- Product
- Market
- Sales/orders (if any)
- Relevant Key Personnel experience
- Funding requirement

It seems a lot to get on one page, but you are only putting down the facts; no flowery bits. The summary should hit them between the eyes in respect of what your company could make over say three years against your required funding; after all, for those who are funding the opportunity, these two are what it is all about as far as they are concerned. Your background should include the IP protection and what stage it is at i.e. filed, granted, GB or European etc. Your market should cite the comprehensive

166

analysis you have undertaken and the potential market size. If you have any advance sales or orders and their potential value, this can do wonders in getting them interested. Concerning personnel, put in there the key attributes of the personnel relevant to the product and the market.

Background
This section is a bit of a catch all (a history, the present and the future) and always overlaps with other sections of the business plan; in a way it explains the processes that you have gone through in this book and is almost an expanded Executive Summary,

- Who the individuals are and their relevant experience
- When and how your idea originated; the POS (problem, objective, solution)
- Detail of your product
- How protection and prototyping were drawn together
- The professional bodies and/or institutions that you employed to get your product to where it is now
- What has been spent to get the product and the opportunity this far
- What if any potential orders you have and who they are from
- How the market has been defined through market analysis
- Where the future lies in the next three to five years for your company and product
- Remember the three legged stool and how you have made solid those three legs to support the seat of the idea to get to where you are now

Product(s)

This section is about the money generator – your product! When you ripped up the rule book some chapters back, you may have arrived at more than one product. Be careful here if you have more than one product, particularly if they are in totally different markets, as it could be seen that you are not concentrating on the primary product that came about through your original thinking, the "must have need". To dilute the focus with too many products is often a fatal mistake from the outset; some of those other products can come later in the life cycle development of your new business. Explain here what types of IP protection you have in place and what stage each one is at.

Probably the most important aspect within this section is the differentiators; those attributes of your product that make it different to others in the market. Why would someone want your product instead of another?

It is appropriate in this section to put in a SWOT analysis on your product. This is Strengths of, Weaknesses in, Opportunities for and Threats to your product. In a way, this could also sit in "Markets and Marketing" section but as it overlaps with the FMEA that you did several months back, there is no reason not to put the SWOT analysis in this section. Whatever you do, don't fudge the Weaknesses and Threats; there is a great temptation to focus on the Strengths and Opportunities as they are the "nice bit". For funding to be forthcoming, it is so important that you have recognised every perceivable weakness and threat and mitigated these in your strategy; in doing so it adds a great deal to your credibility.

Use images of your product that show it to its best advantage, even including images of the packaging that the product will be displayed in if end user packaging is appropriate.

Markets and Marketing

This is where your detailed market analysis kicks in; available market for your product, geographical opportunities, competitors, legislation both current and future and comparison with like products and their success are all important in a convincing argument about the market opportunity. How you will actually go about marketing and promoting your product is incredibly important (as are the costs). It's great to quote market statistics, but if you have no plan as to how the world is going to hear about your product, then forget it. Are you going to market your product on the internet? Or maybe you will employ a marketing company to promote your product or employ a marketing director instead.

Cite the databases (and if you used the British Library, cite them as well) that you have used in coming to the conclusions in this section, it shows some solid work has been done with highly reputable market research and analysis companies/institutions. Don't do this in too much detail; in other words, don't put 30 pages of data and statistics here as you can always elaborate with more information in the appendices at the back of the business plan.

Quoting advance sales and orders or letters of interest in your product (if any) along with the detail of who the potential customers are and their markets must be included.

You will have to state in this section what you believe is the size of the market for your product and how you arrived at such a conclusion. Include the price point (factory gate and retail) and the number of units that can be sold over the next three years to how many potential customers and in which geographical markets.

In this section, if you can support some of your work by using graphs and tables, this helps to give a visual presentation instead of paragraphs of number after number. It also improves the professional look of your business plan.

Manufacturing
So you're going to make your product, but how? The equipment required to do so is dependent on your product. Initially, there are two things to consider; do you acquire all the equipment to make your product from scratch all the way through to its distribution into the market? This can be burdensome in terms of financial commitment. Such an overhead could make the business plan financials unworkable. Or do you outsource some elements of the production and distribution to other companies? The latter is probably preferable as it reduces the burden of capital expenditure in the early part of your new company's life. There is a trade-off to some extent; outsourcing some of the production can mean that part of your product is more expensive to produce than if you produced that element yourself. Conversely, if you did produce that part of the product yourself, you will have to factor the capital expenditure of your new equipment into the cost of the product. Which is the right way to go? Only you can decide along with some sound financial help from an accountant.

Another point to bear in mind is the more equipment that you have, the larger the premises required; more cost!

Manufacturing overseas should be considered if your product is high volume and has to be price competitive to sell. If you have to consider this option and incorporate it in your business plan, don't trawl the internet for overseas manufacturers and contact them yourself. There are many excellent intermediary bodies that can point you in the

170

right direction and on most occasions are happy to make the introduction on your behalf. Chapter Eleven; "The big picture - Assessing the true market potential" gives examples of such intermediaries.

If your product is of a specialist nature, where volume of sales is not great and the price point is not so important as your product is a "must have need", then manufacturing in the UK might be appropriate.

Key Personnel (Management Team)
Incorporate in this section an abbreviation of your CV(s); concentrating on the experience and qualifications accrued that are relevant to the setting up, developing and running of your future business. This applies to each member of the personnel. Start with your most recent and relevant experiences.

What investors are looking for here is skill sets and experience that are complementary to each of the personnel; one may have intimate technical knowledge to make and develop the product, another may have broad day to day operational experience in production, another an in-depth knowledge of marketing and successful past sales in the market sectors appropriate to the product. In other words, the whole of the skill sets and knowledge is greater than the sum of its component parts.

If there is a hole in the skill sets and knowledge of the personnel, make this clear, don't try and cover this void in adapting your abbreviated CVs. Investors would rather you identify and recognise that gap which can be filled by one of the investors as a non-executive director or by recruiting an individual or individuals that will complete the personnel picture. Don't forget to put the overhead of these additional personnel (if needed) in the business plan.

If you are successful in jumping through the first hoops of seeking investment/funding, those that are providing the finance will want to meet all of you. They will ask about the skill sets of each of you and most importantly, they will be watching you; what each of your individual personalities is like and how you interact with each other. If for example, there are three of you, and all are highly energetic, creative people, there is a danger that you can be pulling in different directions. Conversely, if all three of you are "bury your head" types and inward looking at what you're all doing, the company may never get anywhere. They are looking for a character balance; the creative one, the "roll your sleeves up and get it done" type and finally the rock-steady type who can steady the ship and pull all the threads together when required. Yes, this is a theoretical example, but it gives you a feel of what they will be looking for.

Investors consider the Key Personnel almost as importantly as the product and the business plan proposal in its entirety; after all they are investing in you and trusting you to manage and grow their money wisely and diligently.

Funding Requirements
This is driven by the cash flow requirements when you have considered sales, cost of sales, overheads and when they occur in the three year business plan. Cash flow forecasts come after this section. The total funding requirements will depend on what type of funding is required, so this section of your business plan will change depending on who you are approaching.

If you are seeking funding from an investor in exchange for a percentage (equity) of your company, then apart from a possible funding set up fee at the beginning, you will not normally have to incorporate any repayments to them in the cash flow forecast. If you are seeking a loan from a bank or financial institution, then regular repayments of the loan will have to be incorporated into the cash flow forecast.

In its simplest terms, it is the difference between these two types of funding that will define from your cash flow forecast the level of funding required depending on the funding sought. It can get more complicated as some companies have a combination of loan and equity funding. Advice as to which is best for you should be sought from an accountant.

The total amount of funding required is usually based on the accrued monthly losses that occur from day one when set-up and running costs exceed income. If everything goes well, as sales and revenues start to kick in, the monthly losses stop when monthly income exceeds monthly outgoings – this doesn't mean profit at this stage, it just means that more money is coming in than going out (positive cash flow). This is a very basic way to look at the cash requirements as there are many other factors that can affect cash flow, but is a thumbnail sketch as to how to consider your total funding requirement.

When you have realised your funding requirement, put it in this section along with how long it is required for. This will be the point when your business plan starts to show a monthly positive cash flow.

Schedule of overheads
This is in effect a list of yearly overheads required to set up, run and develop your company. This includes staff, establishment costs such as premises, capital equipment, marketing, running costs of premises, loan repayments (if applicable) and many more, smaller costs that regularly or occasionally occur. Your accountant will be well versed in

all the costs that should be considered along with the frequency of their occurrence.

Set out these overheads for years one, two and three.

Schedule of income
This could be considered the make or break part of your business plan. If the forecasted sales are too low or if sales do look good, yet the profit margin after the cost of sales is not good enough, then on both accounts, the business could be in trouble from the early days.

Set out your annual sales along with the annual cost of sales on a per annum basis over three years; the income is the net sum of these two figures (gross profit). Show this figure also as a percentage.

Show here as to how the sales are arrived at, how many customers you either have or will have and as to what sort of payment terms you are allowing those customers.

Also show how many suppliers you are dealing with and the credit arrangements made with them.

Over the three years, the sum of your gross profit less the sum of the overheads for the same period will show your funding requirement

As with the schedule of overheads, set these sales out for years one, two and three.

Three year cash flow forecast
This is where the work of the trusty accountant comes into play. A point to note is when putting the cash flow forecasts together, take into consideration any credit

arrangements (the payment terms) that you may give to clients; do the same with any credit arrangements that your suppliers may give to you. It is the cash flow that dictates the funding requirement.

Once you have worked out your three year cash flow forecast and are satisfied that it best represents the potential of the business, do two more; one that represents an increase in sales of 50% and one that represents a decrease in sales of 50%. These do not go in to the business plan, but they give you an idea of what should happen if sales turn one way or the other and helps to equip you when questions are asked when meeting up with possible investors or finance institutions.

Three year profit and loss forecast
The profit and loss forecast will be driven by the three year cash flow forecast. In each of the three years, it will show whether your proposal is making a profit or a loss. It is not uncommon for businesses to make a loss in the first year, primarily due to the set-up costs and the period of time it takes to generate a substantive, sustainable market. If your business fails to make a profit after three years, it is very unlikely that finance for the proposal will be forthcoming. Similarly, if the proposal forecasts massive profits at the end of year three, there might by eyebrows raised – it could appear that you have been far too optimistic with your forecasts. Investors and financial institutions like to see a healthy profit at the end of year three that is represented by a steady, strategic growth over the preceding two years.

Three year balance sheet forecast
Once again, this is the accountant's forte and demonstrates the liquidity of your new business at the ends of year one,

two and three. It will show how solvent the proposition is and helps demonstrate the viability of your proposition. If the balance sheet that has been worked out shows that the proposition appears insolvent, then along with your accountant, you will have to readdress the numbers to make the proposition viable.

Appendices
Use appendices when further information throughout your business plan is required. Don't crowd out every section with every piece of information under the sun; it is better to be punchy and explicit in each section. When there is more information available, put in the relevant section, for example "Further information can be found in Appendix B – European Market Breakdown" or another example, "The full patent application can be found in Appendix E – GB Patent Application".

Seeking funding from banks and institutions
Earlier on in the development of your idea, you may have given away a percentage of your idea so that your idea was developed through to a final prototype. It might be for this reason that you do not want to seek another equity investor, so diluting your percentage even further. It maybe that you never wanted to give any percentage away from the outset, or you have sought investment in exchange for equity and have not been successful. Whatever the reasons behind your decision, you have decided that a loan is the route along which you want to go. The Small Firms Loan Guarantee Scheme, which was mentioned previously, may be a possibility, but certain criteria have to be met; speak to either your bank or accountant regarding details of the scheme. For a straight forward loan, you will have to provide some form of security; homes being the usual security against such a loan. You will find that with banks

and financial institutions the process of obtaining a loan is nowhere near as complex or challenging as seeking an equity investor in exchange for a percentage of your business, but a robust business plan and evidence of the ability to repay the loan is essential.

Seeking funding from investors
Armed with a complete and viable business plan, if you don't want to go along the loan route, it is now time to approach investors and/or finance institutions. Approach them with the one page Executive Summary. Your accountant or an intermediary such as Business Link will be able to advise you who to approach, for example business angel consortia or High Net-Worth Individuals.

Should your executive summary attract interest, they will then wish to have a copy of the business plan. If, once having pored over that, their interest has not waned, it will then be time for you to meet up with them. This meeting could be on a one-on-one basis or could be with several of them at the same time, whichever is the case, know that business plan inside out! If there are two or more of you, make sure that each of you is well versed in the specialism that each has relating to the business plan, although that is not enough in itself; each of you still has to have a total understanding of the overall business plan.

An alternative to approaching investors directly is "showcasing". There are various organisations and intermediaries who at certain times of the year have funding showcase events. If you apply to such organisations with your executive summary and are successful, you will have the opportunity at such an event to "pitch" your proposal (anything from 5-15 minutes) in front of a group of potential investors. The benefit here is that you are killing several birds with one stone. If you are unsuccessful, don't dishearten as the experience gained from you presenting and from them asking a plethora of questions will help you polish up your presentation to other investors.

When presenting at such events, your opening gambit is a verbal form of your executive summary (including the funding that you are looking for and the anticipated profit at the end of year three) followed by more detail about your product and the market for such. Mentioning that you have got orders (if you have) for your product is a big hook. Your experience relative to the proposal is important in warranting such an investment and if there are two or more of you, get it well rehearsed as to who is saying what and when. Will you be nervous? Of course, but so will others who are presenting and whether you're first, last or somewhere in between in the presentation pecking order, do not worry about how others present or what their product is – it's all about you, your product and your proposal.

Whichever route you take, hopefully you have finally got some investors interested – Whoopee! You are now closing in on a deal which if concluded, will fund you and your proposal for up to the next three years, but the work is not finished yet. Investors will want to know what level of equity you are prepared to give them for their investment. Don't go down the route of offering them a ridiculously low percentage as one, it shows that you are trying to have your cake and eat it and two, it's a bit of an insult to their intelligence. For them to invest in what appears to be a possible viable proposal, you would expect to give away in the region of 25-40%. That may appear a lot, particularly if there are say three of you seeking the funding together for the proposal; it means that you may only end up with 20% each, yet if say the funding is £500,000 and your business plan shows a potential profit of say £5 million in three years, then if that was given out as a dividend, your dividend would be £1 million (dependent on the type of shares issued) – not bad for three years work! This doesn't take into account the value of your shareholding should you sell your shares at some point in the future.

Never concede more than 49% of the equity in your proposal as you do not want to lose control. It is very rare for investors to require more than that as they appreciate that you will be running the business and after you have put in all that hard work and effort to get to where you intend to be, losing control of your business is somewhat of a disincentive.

If you are successful in sourcing an investor, one condition (amongst many) that they will stipulate is that you have to put some money in as well. It will be nowhere near the amount of money that they are putting and it may only be a nominal amount, but it will be at a level that shows a high level of commitment by you and locks you in to make sure that you put every effort in to make the proposal work.

So the investors have backed you, they're putting in a sum of money to support the cash flow of the business in exchange for a percentage of your business. They also have a non-executive director on board to help and advise you where relevant on the successful development of your business. A point to note is, the non-executive director is also there to see that you spend their money wisely!

So now the hard work starts; your idea which was only a notion say two years ago has robust IP protection and the pre-production prototypes are now going into production as the actual product to fill the first order book. Whatever you do, commit all your time and effort to your business; don't over-focus on one area to the detriment of other parts of the business. Maintain open communication with your colleagues – its better to air any concerns as soon as they arise than carry them whilst you're busy. Visit your business plan regularly to see that you are on-track – no one has ever got a business plan 100% right, it just doesn't happen, so if you deviate from the plan don't get too concerned. The important thing is that you mitigate the risk of going entirely off-track by identifying corrective measures to bring you back in-line as early as possible.

There's an old adage that springs to mind – "he that sows, reaps" This is probably the only opportunity that you will have of funding, so give it everything you've got over the next two or three years; sow those seeds of opportunity and effort and hopefully, you will be able to reap a rich harvest in the years to come.

Chapter Thirteen

Licensing, assigning and evaluation

So manufacturing and selling your product yourself is not the option you wish to take, maybe because the financial investment to do so is too large a risk or your experience in your particular field is not appropriate to do so. There could also be many other reasons why taking this route is the least preferred.

The other option is to take out all the risk in commercialising your product yourself and licence it to another party. In doing so, the other party takes on the risk of making it a success whilst paying you for the privilege of doing so.

What is a licence?

Simply, a licence is a binding agreement between the owner(s) of the IP to be licensed and another party who wish to have wholly or part exclusive rights to manufacturing and/or selling the IP for commercial gain. Once in force, a licence is a legal document that binds both parties to the terms contained within. It is normally written in a way that means it is a win-win for both parties.

Whilst a licence allows the licensee to operate within the terms of the licence, often with a degree of freedom, it conversely will restrict you from possibly seeking other commercial opportunities, so you have to be sure that the licence agreement that you enter into is right for you; exclusive to the licensee or non-exclusive, the latter allowing you to seek other licensing opportunities.

How long is a licence for?

There is no hard, fast rule here – factors that affect the length of a licence agreement are manyfold,

- The level of investment to make and market the product by the licensee; the higher the investment, the longer the licence period typically

- Size of market; the bigger the market, again the longer the licence period

- Stability of the market; the more stable the market, then a longer period for the licence will probably be sought

- Strength of the IP protection; if it is incredibly strong, has been examined and in the case of patents, all objections have been successfully contested, then this will more than likely warrant a longer period for a licence agreement

- Geographical extent of the licence; the more international the agreement, the longer the period of the licence tends to be

It really is up to you, what you feel is right for you and the licensee. A tip here; it may seem great to have a licensee tied up for the life of your patent i.e. 20 years, but what if, after a year or so, the manufacturing and sales dry up? You are then tied into many years in the licensing wilderness.

I suggest that a licence agreement of five years is often an ideal period; it has proved so in the past. This way if the licensee fails to perform in respect of manufacturing and selling your product during the period of the agreement, at least you haven't got another fifteen years on top of that before you're free to do what you want.

This period of five years should however, carry various conditions:

- You have the option to extend or withdraw the licence from the licensee at the end of the licence period, giving say 12 months notice (i.e. at the beginning of year four)

- If the licence is to be renewed, then the licensee has first option to renew but on the payment to you of a renewal fee set out at the beginning of the agreement

- If the licensee breaches any terms and conditions of the licence agreement, then you have the right to withdraw the licence immediately and are free to seek a licence agreement elsewhere

Five years is a reasonable length of time for a licence agreement. It permits the licensee development time to produce and build up a market for your product (up to 18 months) and in doing so, gives them enough time in which to reap profitable revenues (after all that is why they entered a licence agreement in the first place). If the licensee performs well beyond expectations, with considerable growth to come, then you can always open up negotiations at any point to extend the period of the licence from say five years to ten. Not only does this secure income for you for possibly another five years, but it offers the licensee security in the knowledge that they can continue to commit money and resource for greater returns over a longer period.

Coverage of a licence agreement

What is the coverage of the licence in geographical and industrial terms? Again the choice is yours. Your market analysis will have told you where the market lies for your product and along with your "outside the box" thinking have identified a priority list of uses. The list of potential licensees will also help scope the coverage of your licence agreements.

It will not become totally clear until entering licensing negotiations with a licensee or licensees as to what the coverage of the agreement will be as they will have their own ideas if they are interested in your product, but it helps to be forearmed. A few simple things to think about:

- Does the potential licensee have a regional, national or international market? If international, which countries?
- Are they the biggest player in the market or are they one of many?
- Do they cover different industry classifications i.e. domestic goods and industrial goods?
- Have they any partnerships with, or ownerships of other companies?

If on the largest scale, you were to licence your product to one major global corporation whose distribution covers three quarters of the globe, it would probably be best give them exclusivity around the world (it is something that they would probably insist upon). In doing so you've only got one licensee to deal with and it is far easier to keep track of things when only dealing with one. The downside is that you have closed out any opportunity to licence elsewhere – all your eggs are in one basket. If sales go well around the world, you are indeed sitting pretty comfortably, but beware of such large organisations – they could quite easily drop your product if another more lucrative product comes along. They might have to write off tooling and marketing costs into the hundreds of thousands or even millions of pounds, but that is small beer to them. You will be protected to some extent by what are called Minimum Performance Guarantees (MPGs) or Minimum Volume Guarantees (MVGs) and these are something which we will go into shortly, but by having licensed to such an organisation, the true potential of your product may never be realised if such aforementioned events occur.

The opposite end of the spectrum is the smaller, localised licence agreements with two or more licensees. Yes, getting multiple licensees requires a great deal of work, but at least all your eggs are not in one basket. Your market analysis may show that there are a couple of potential licensees in the USA, one in the UK and maybe three or four in Europe. In an ideal world, you would licence your product to one in each of these geographical areas, giving you three licence agreements; each having exclusivity in its respective country, or in the case of Europe, countries. That way, your options are spread and if one fails to perform, you can withdraw that licence and seek an alternative licensee. Whilst you are doing so, you still have two other licensees "firing on all cylinders".

If you have identified say, two uses for your product, each being a "must have need", then you can look to licence your idea to two licensees in the same country so long as they are in different industry sectors and are not competing with each other in the same market. An example could be a new type of electric brush where you have identified two uses; one being a toothbrush, the other being a cleaning brush for car valeting. One use could be licensed to an oral hygiene company, the other to an automobile accessory company. Both licensees would be permitted to sell their respective product in their specific markets only and it would be written into the licence agreements that they would not be allowed to sell into each other's markets.

Another possibility in multiple licensing is where the company that you have identified has international partnerships with other companies. In this instance, you may wish to licence to the main licensee and through their partnerships, grant sub-licences to their partners with the same terms of the main licence applying. In this way, through your negotiations with the main licensee, you have increased the distribution of your product by granting such sub-licences.

What can I expect to earn from a licence agreement?

Again, it really is dependent on your product, the market size and the price of the product. Along with these, what you can earn is dictated by,

- The initial licence fee that is paid to you by the licensee for the rights to the licence

- The royalty that is paid to you by the licensee for each unit manufactured or sold (there is a subtle difference between the two and we will come back to that)

- The period of the licence

- Licence renewal fees

But before we delve into these points, there is another alternative; assigning the rights in your product to another party, called an assignee. Assigning rights is useful if it suits you to have a lump sum of money instead of waiting for royalties to come through. This can be,

- From an initial one-off payment; then no further payments made by the assignee. This is an absolute assignment, transferring ownership of your product's IP to the assignee

- Through an annual fee paid by the assignee over an agreed period of the assignment

Both of these options mean that no royalties are paid to you by the assignee. One distinct advantage is if the assignee fails to perform in terms of the anticipated numbers of units sold of your product, then if you had taken the royalty route in a licence agreement, you could have been worse off. Conversely, if you opt for the one-off or annual payment options as stated above, you could lose out on high royalty revenues if sales of your product exceed all expectations.

So which do you choose? Only you can decide and your market analysis should assist you in your decision making. Your particular personal needs as well may help swing the decision one way or another. Another influence is of course the potential licensee; they may want to go along the royalty route, but they may prefer to assign your product's IP.

So irrespective of which way you go, how do you work out what to ask for? In its simplest of terms, you ask for what you can get away with and it's dependent of course on what the potential licensee is prepared to pay.

Initial licence fee
When it comes to the initial licence fee if you're going down the royalty route, you should expect that fee to be at least the amount that you have spent on getting your product to where it is. This is reasonable insofar as the IP protection and prototype development along with CAD files that you

187

have would generally have cost the licensee at least that much if they had done the work themselves. So if you have spent £20,000, then expect at least £20,000 as an initial licence fee. You could probably ask for more and that is dependent on what further work the licensee has to do to get the product produced and promoted and as to how big they see the potential for sales over the period of a licence agreement.

On a hypothetical basis, if the tooling and set up to make the product costs say £100,000 and sales over an agreement of five years could net the licensee £750,000, then asking for £50,000 for an initial licence could be a possibility. Generally, the higher the tooling and set up costs to get your product into production, the more you can ask for in terms of an initial licensing fee; it shows that if they are prepared to spend considerable sums to do so, then you have them hooked. They would not spend so much in getting your idea out of the blocks if they didn't believe it had real potential.

In licensing negotiations, there is a fine line between getting away with what you can and being greedy which may get the licensee's back up resulting in rejection. There's a difference between what you need and what you would like; push the boundaries of reason when asking for what you want, but don't step beyond them.

An up-front assignment fee is for want of a better word a "take the money and run" one off payment. This has to be gauged against what you could earn through the licensing route and it is a big commitment by an assignee to pay for such. It should be balanced against what you could earn in a licence agreement over say five years against the benefit of having a lump sum up front. A sum in the region of around 50% of potential earnings over a five year licence agreement could be considered reasonable (after all, the assignee is taking quite a risk) and whilst not as much for you as if it was a five year licence agreement, it is guaranteed and in the bank. This is not a hard and fast rule and is down to what you personally want from such an exercise and what you feel you could obtain.

Royalties

One of the big questions regarding the level of royalty payable per unit sold is what should that level be? There are many different opinions on this, but I've worked on a sum representative of 10% of the factory gate price being a good benchmark and in negotiations, this has generally never been far off the mark. If necessary, you can always drop that percentage down, but you can't ramp it up once you've shown your hand. There are many variables applying to this; type of product, cost to produce and market size. The lowest percentage royalty that I have been involved in was 5%; it was for a low priced product that was in an extremely price competitive market indeed and to ask for more would have made the product uncompetitive, but the potential volume of sales was very high, so making the rewards very good indeed. The highest royalty was 25% for an expensive specialist product having unique attributes that once purchased by the end user offered them excellent medium to long term savings. In doing so, the product could be sold at a higher price in the market to similar products, so allowing for the margin of the royalty.

One easy mistake when trying to calculate what sort of income you could possible earn is calculating royalties on the retail price of the product not the factory gate price. If your licensee is a manufacturer or possibly a distributor, then the royalty paid will nearly always be the factory gate price.

When agreeing the level of the royalties, it is important that royalties are paid on the number of units manufactured, not on the number of units sold or when the manufacturer is paid. You will come up against some opposition to this, but the reason is fairly straight forward; manufacturers may make the product, but may not sell it for several months. It may then take three months before they are paid, in total some six months after they have manufactured it. The suppliers that supply the manufacturers with the raw materials are paid when the raw materials are supplied or certainly within the terms of credit offered by the suppliers; they don't wait until the manufacturer has sold the products. Similarly, you are no different to a supplier; you are supplying them with your IP and the royalty along with the other raw material costs is factored into the manufacturing of the product. You should therefore not be treated any differently. After all your IP is like a raw material; if you along with other suppliers hadn't supplied them in the first place, the licensee couldn't make the product.

I also came across an occasion where a licence agreement stated that royalties were paid on the sales of a product. A licensee manufacturer made several thousand units of a product and then gave them away free as part of a promotional exercise. Agreed it was to promote sales of the product, but the royalty along with the other costs of the product were included in the marketing department's budget for the promotion and therefore could be argued that royalties were due to the licensor. The agreement for paying royalties was not based on units manufactured, but on those that were sold. The licensor (IP owner) subsequently relinquished any rights to royalties on those thousands of "freebie" units given away.

Another point to ponder is do you agree a sum per unit sold for your royalties or a percentage of the factory gate price?

If the sum is a percentage, then your royalty can go up or down per unit depending on the factory gate price. I have found that using the percentage as a negotiating point to arrive at a fixed sum per unit is a useful start. For your future planning purposes, it is often better to agree a fixed sum per unit which remains constant irrespective of factory gate price. If for example, the factory gate price of your product is £3.50, then if you agree a royalty of say £0.35 per unit, it will stay constant regardless of whether the price goes up to £3.70 or down to £3.20.

Renewal fees

Renewal fees only apply if there is a licence worth renewing. You may have given notice that you wish to cancel the agreement if the licensee is not performing well enough and be looking for another licensee to give your product a new lease of life. Conversely, the licensee may also wish to give notice of termination. If on the positive side, the licensee is performing well, then a renewal of the licence agreement would be the natural thing to do in the interest of both parties. It is in effect the granting of a new licence, so you can either have the renewal fee written in to the licensing agreement as a sum from the outset or have a clause that states that the amount of the renewal fee will be agreed by both parties say twelve months prior to the end of the agreement. This can be risky as if there is a failure to agree by the two parties, the future of the licence could be in jeopardy. Of the two possibilities, the former is preferred as each party knows what the renewal fee is from the outset. There is no reason not to set the renewal fee at the same amount as the initial licensing fee. You may however be able to get more if the product is selling really well.

Minimum Performance Guarantees (MPGs)

Sometimes called Minimum Volume Guarantees, MPGs are guarantees that the licensee will pay royalties to you on a minimum number of units manufactured per year, irrespective if they do not achieve such volumes. This has the advantage for you of being paid a minimum sum per annum by the licensee. Some potential licensees baulk at the idea of this, but if they are serious about licensing your product, they will have done their sums as to the sort of volumes that they believe they can sell and armed with this, an MPG level can be written into the agreement. Another reason why a guarantee of this sort should be written into the agreement is that payment for such will be coming out of one or more departments' budget within the licensing company. If they have an annual commitment to paying out to the licensor (you in this case), there is no better incentive for those departments to try and make sales of your product exceed such guaranteed levels.

If a new product is presented to your licensee and there are no MPGs within the agreement, they could drop your product like a hot potato and pursue the new product with total disregard to yours. Yes, they will have paid you an initial licence fee, but if the commercial potential in the new product that has come along warrants writing that off, you are left holding what is in effect a worthless agreement in terms of future royalties.

If a prospective licensee does not agree to MPGs, then don't enter into a licensing agreement as you could be waiting on a wing and a prayer for any royalties to come through. Move onto the next licensing opportunity.

So at what levels should the MPGs be set? Again, it's a bit of a "suck it and see" exercise, but it will become clearer once you have had at least a couple of meetings with the potential licensee. You will have ideas of how many units can be sold per year and so will the potential licensee. If you are both in agreement as to these possible numbers, then as a rough indication, MPG levels should be set at around half or two thirds of the anticipated annual sales. MPG levels are not discussed at all until you have found out what sort of sales levels the potential licensee might achieve (this will be discussed in "Presenting to potential licensees" later in this chapter). MPGs in year one of the agreement should be set at a lower level because of the lead-in time required to tool up and initially market your product. A more realistic level would be one quarter of possible ongoing annual sales as a level for the MPGs in that first year. The ensuing years of the agreement should have an increase each year in the MPGs. You will have to negotiate this, but 20%-30% increase per annum is not unreasonable.

What if the annual sales of your product exceed the MPG levels? Then the additional royalty revenue will be paid to you. Normally, the MPGs will be paid to you at intervals of two months up to six months, depending on what you both agree. At the end of each year, the sales are reconciled and then if sales do exceed the MPGs, a balancing royalty is paid to you for the difference between the MPG revenues and the actual royalty revenues for that year.

Evaluation Period

On occasions, a licensing opportunity could be missed if the potential licensee feels that there is a lot more development work to do on your product to suit their product profile and wish to undertake their own extensive market analysis, particularly if your product is in a different market to their traditionally established ones. To commit to either a licensing agreement or an assignment may be too heavy a commitment if there is that level of uncertainty attached to your idea, despite the fact they very much like what you are offering.

An answer to such a scenario is to offer them an evaluation period which is the precursor to a possible licensing agreement or assignment. So how does this work? An evaluation agreement is very much a short term licence in effect; it gives the potential licensee the opportunity to have exclusive evaluation of your product, typically for a period of three months at the end of which they either wish to licence your product, decline it and return all the IP in the product back to you or request for an extension of the period of a month or more.

An evaluation period agreement should where possible contain at least the following,

- The length of the evaluation period (normally a minimum of three months)

- Option to extend the evaluation period if required by the potential licensee (typically one month increments)

- The fee for the initial period which is non-refundable and payable up front

- The fee for any extension to the period (normally a pro-rate of the three month period) and payable at the beginning of the extension

- A description of the intellectual property

- An undertaking by the potential licensee that should they decline the opportunity at the end of the evaluation period, they will return all the IP along with any improvements made to such

194

- The potential licensee relinquishes all rights to pursue any further interest in the product

For the licensee, an evaluation period is a damage limitation exercise in terms of risking a moderate sum of money to evaluate an opportunity without committing to a licence from the outset.

You cannot really lose pursuing this option and it can have distinct advantages for you; first you're getting an up-front fee, maybe as much as £10,000 or even more for the evaluation period and if the licensee wishes to take the product on at the end of the evaluation period, they are then more certain than ever that they possibly have a winner on their hands. This can help drive up the numbers when it comes to you negotiating the licence fee and royalties or assignment fee.

If they decline to take up the option of a licence agreement at the end of the evaluation period, then again, not only have you received a non-refundable fee, but they have returned all the IP to you along with any improvements that they may have made. In effect, you've been paid by a company who has evaluated your IP at their expense and time, possibly having improved it along the way! This leaves you free to seek another licensee or assignee or possibly another evaluation period elsewhere.

Other points to consider in a licence agreement

There are many other points to include in a licence agreement apart from those described above and those listed below are not exhaustive. Always seek professional help when drafting a licence agreement. This help won't come cheap, but if you are at the point of drafting a licence agreement, you are most likely to have a verbal agreement from the licensee to licence your IP, therefore the professional costs should dwarf what income you are going to receive over the coming years from fees and/or royalties. Other points to go into the agreement,

- The legal names of both parties entering into the agreement along with, where applicable, registered addresses and company registration numbers

- Full description of your IP and any patent, registered design and trade mark numbers including geographical regions of coverage

- Maximum factory gate price at which units of your IP will be sold; you do not want the licensee to have total freedom to sell units of your IP at a higher price and higher gross profit margin which could seriously affect its sales potential

- The licensee will maintain any renewal fees that become due on the IP protection or the licensee will pay you for the renewal fees and you renew accordingly. The latter here may be the preferred option as if the licensee fails to pay renewal fees, then your IP protection could lapse and render it potentially worthless. If you are responsible for the renewal fees, then you are in control

- The licensee will ensure that any laws and legislation that are applicable in the manufacturing of or distribution of your IP are adhered to

- The licensee wholly indemnifies you against any liabilities that may occur from the manufacture of, distribution of and use of your IP

- The licensee will pay MPGs at intervals of two months from the beginning of the licence agreement (you may agree that this interval is every three months or even six monthly) and provide you with a written report from the licensee of units manufactured in that period

- After twelve months from the beginning of the licence agreement (and at the end of subsequent years after that) the licensee will provide an annual report of how many units have been manufactured in that year together with a top-up payment to you if the MPG levels have been exceeded

- If the licensee sells their business in part or whole, you have the option to cancel the licence agreement if it suits you and be paid any outstanding money

- If the licensee acts in a way that is neither legal or ethical, you again have the right to cancel the licence agreement

As you are probably using a professional to draft your licence agreement, they will most likely have additional material to be included, and they will write some of the points above in a more legalistic way.

Contacting potential licensees

Armed with your detailed market analysis, you will have identified companies that could be potential licensees of your IP together with contact names; so who do you write to? Generally if the company is an SME (under 250 employees), then go to the top; write to the Managing Director or Chief Executive; the company is small enough for your letter to reach such a position and be considered. It may be passed on to a relevant individual within the company, but the fact that it has been passed down, means it will be acted upon. If the company is larger, they may have a department whose responsibility includes considering new and novel ideas; it would be to this department that your letter should be addressed.

Keep your letter to one page with a page of marketing statistics attached. You only get one chance with this letter, so make it relevant and punchy. Include,

- A brief description of your product, (keep it brief and don't breach any public domain issues as you do not have a confidentiality agreement in place at this stage)

- The stage that your IP protection is at i.e. UK patent filed or PCT filed etc

- The market that your product is appealing to

- That your product complements the product range of the company

- Your intention to licence your product and that their company fits the profile that you are looking for

- A line referring to the attached market statistics

- That your product could possibly make a significant bottom line contribution to the company's balance sheet (this is invariably a big plus to include this as an executive can hardly ignore an opportunity to assess a product that would make such a difference)

- A reference to you contacting them in the next few weeks

Your market statistics on the attached page should hit all the buttons relevant to that company. If for example, the company only operates and sells in the UK, too much detail on worldwide statistics would be irrelevant, but it is worth making brief reference to them as they may indicate an opportunity that the company had not considered before. In this instance, detailed market information on the UK market along with competitors' market share would be useful to show that you know what you're talking about; it also shows that you are aware of their competitors and in a very subtle way, it is saying to the company that those competitors are also potential licensees.

There is a temptation to write to every company that you have identified from the outset, but don't; only write to four or five. The reason for this is some of those companies will reply or you will eventually contact them yourself and even if all of them decline your product at this early stage, the reasons they give as to why they declined could be very useful information in reshaping your letters to other potential licensees.

If after around three weeks, you have not heard anything from your initial letters, then that is the time to try and contact them by phone. It is important to note that if you are writing to the larger corporations for example, those with a global presence, they may take up to three months to reply to your initial letter. Some larger corporations may come back within a couple of weeks of receiving your initial letter with a submission form and confidentiality agreement, both of which have to be completed and submitted for consideration. Again once having done so, you may not hear anything for up to three months. Some of these corporations will not consider a submission unless a patent application has been filed in the countries relevant to their market and they will ask for the relevant patent application numbers to be provided.

If you are unable to contact an individual by phone, send a second letter, including a copy of your first letter with statistics attached. In the second letter, make referral to the first letter and politely request a reply.

If the second letter doesn't work, send a third and final letter stating you have not heard anything and you presume that they have no interest in your product and because of its potential you will seek a licensee elsewhere. Thank them for their interest. Sometimes a third letter like this may just evoke a response.

All this effort can take up to two months (larger corporations excepted) and can seem like a long slog, but sticking to the disciplines could invariably result in at least one response requesting a meeting.

Preparing a presentation to a potential licensee

The letter finally arrives requesting a meeting; the interest is there, but the fish is far from being firmly caught on the hook at this stage. You will normally be given half an hour to do your pitch; at this stage they have an idea of what your product might be, but they do not know what it is in its entirety. It is this half an hour that is a "do or die" opportunity, so prepare for it well. It might be a meeting with the person to whom you wrote, or there could be two or three in the meeting, for example a Chief Executive, a product designer and possibly someone from marketing.

As when you met your patent attorney for the first time, prepare your presentation well and it should include,

- How the idea came about

- A copy of your IP protection i.e. patent and or registered design application

- A power point presentation showing why you chose them as a potential licensee, (this always goes down well!) along with market statistics relevant to that particular company. If not a power point, then a presentation that is printed out

- Your prototype(s)

Equipped with these, you are ready for that first meeting.

Presenting to a potential licensee

First and foremost, before you go to the meeting, ensure that a confidentiality agreement has been signed between you and the company, if not prior to the meeting, then on the day itself prior to discussing anything.

Make sure that you are dressed smartly for your meeting; it is incredibly important as a smart suit or a dress suit comes across far better than a jumper and jeans; it invariably puts you on a par with those that you are meeting. In the first ten or fifteen seconds of your meeting, they will have made a subconscious judgement about you and your intent, so being smart and relaxed goes a long way to getting them on your side from the outset. Do not be nervous; if you are, you are at a distinct disadvantage and it could affect the outcome of the licensing opportunity. I know it's very easy to say, but just treat those individuals as flesh and blood just like you. You will often be pleasantly surprised that your presentation shows to them that you know more than they. That sounds bizarre as they are running their company, but believe me it sometimes is the case.

You are presenting, so you are in control of the meeting. Present in the order shown in "preparing a presentation to a licensee", giving some information on the background to the product, explaining your IP protection, show the power point and comment accordingly or if using handouts, pass them around. Finally once that is complete, reveal the prototype – this often makes a real difference to the level of interest. It's real, they can touch it, poke it and see it in all its glory, imagining what it may look like on a display shelf or even in the users hands.

Throughout the meeting, their brains will be whirring with the one thought coursing through their veins – is this going to make us money?

The first meeting has a two way purpose, for them to assess their interest in your product, but just as importantly for you to assess them; are they the right company to licence your product? It is important to remember this; do so and it puts you on a par with those to whom you are presenting. Ok, you're desperate to get a licensing deal, but not at any cost and the first rule is, he who talks first, loses. This is in respect of money. In this first meeting, money should not be discussed; if they ask how much you are looking for, reply that at this stage you are not here to talk money, but to look at the opportunities your product presents.

If there is obvious interest from the company, the meeting will go on longer than half an hour, but it should not go on longer than around an hour. If it does, there is a danger of you overselling yourself and the impact of your presentation will be reduced. If you have got this far, it is unusual for a company to say no to licensing your product at this very first stage, but it does occur. The outcome is usually a request to hold onto the prototype along with the information that you have provided; this is quite normal to do so. They will then run it by their marketing people, their product development team along with a couple from the production side. They may well say no for a variety of reasons and return everything to you. Don't be despondent; use the experience to gain further confidence should another opportunity come along to licence your product elsewhere.

If the outcome of your first meeting is positive, then a second meeting will be called to discuss either further technical issues (this is where an evaluation agreement might be appropriate) together with a very basic framework for the licensing opportunity. If it is the latter, to get started on drafting an agreement, you will need to discuss and hopefully agree,

- If it is an assignment
- If it is a licensing agreement
- What is the up-front fee?
- What level the royalties may be
- The geographical region
- If the agreement is exclusive to them (you may want to licence to other companies in different countries)
- Length of agreement

Mention in addition that you will be requiring minimum performance guarantees as part of the deal – do it at this stage as you have the fish now firmly on the hook, although beware, it could still wriggle off. The levels for this can be agreed over the ensuing weeks.

Do tell them what you are looking for in these main points of the agreement and take a positive approach to this; don't make asking for an up-front fee and the royalty level seem like an apology. There will be several elements of the framework that they will not necessarily agree with, after all, it is business and normally a compromise between what you are looking for and what they want is agreed upon. If you're not sure, don't agree there and then; tell them that you'll consult your advisors and get back to them – they will probably want to do the same as well. Eventually after many phone calls and emails, both you and the company will agree the main elements of the framework for the licensing agreement or assignment.

If this is the case, then once agreed, the minutiae of the licence agreement can be set out and agreed between your advisor and the company. Once the T's are crossed and the I's are dotted, both parties can sign the agreement, and congratulations, you're there. Writing it here seems all so simple, but you will have to hold your nerve as it is a lot of hard and diligent work, particularly if there is travelling involved. From that first presentation to signing an assignment or licensing agreement, three or four months could have passed, but if it is all signed up, then congratulations; sit back and wait for the cheques to drop through the front door!

Chapter Fourteen

A case study - Walkodile®, a story of success

Elaine Stephen and schoolchildren with Walkodile®

Elaine Stephen, a primary school teacher from Peterhead in Scotland, found that walking along with half a dozen of her class outside of school was sometimes not that easy. The children, in the three to seven year old age group had a tendency to wander or not look where they were going. To keep them all together safely was a considerable responsibility. Elaine explains,

"When parents leave their child with me at the school gates in the morning, they're giving me the most precious thing in their lives to look after. It's a tremendous responsibility.

"Young children are very unpredictable and no matter how careful I am and how much adult supervision I have with me, there's always the chance an accident could happen. I wanted a product which would make the children safer. There wasn't one, so I created my own."

That was Elaine's Eureka moment; what if they could be safely linked together and then she, or any other teacher could take control of the harness and guide the children safely along?

Elaine, along with her husband John set about making a basic working model, employing limited help along the way. They made one mistake though; they went straight for the filing of a GB patent although Elaine's idea was still evolving. They encountered considerable difficulty in getting the working model and first prototype to work as was needed. Their product which they called Walkodile® (note the registered symbol as they went for a trade mark filing as well as a patent filing) was now starting to cost considerable time and money. Twelve months had nearly passed since the filing of their patent application so they had to make a decision about extending the coverage of their application and they decided to file a PCT patent application giving a priority date around the world back to their original GB filing date.

All this was fine, but Elaine and John were getting into a serious level of financial commitment; so far the development and protection had cost them in the region of £50,000 and they still had not got a prototype that was working properly. John commented,

"It was just not coming together properly; we were now into something in which we had no experience. It was a different league to what we were used to and we'd already cashed in insurance policies and re-mortgaged our house to get to where we were at that stage; a working model that just was not the ticket!"

John had seen De Montfort University on the BBC's Innovation Nation television series and decided to contact them, possibly a last throw of the dice as far as Walkodile® was concerned. Peter Ford who ran the development of the project under the Improving Business by Design funded programme adds,

"John and Elaine had a wonderful idea, but the stage where they had got to was racked with problems; their product was far from being production ready, in fact it was no where near being a final prototype."

The Improving Business by Design initiative certainly helped Elaine and John on the cash side of things as De Montfort University was able to part-fund the development through to final prototype; however Elaine and John still had to part with more money. They had been at the point of throwing in the towel, but decided to persevere in the last chance saloon.

In the following months, after several meetings and tweaks, a production ready prototype was finally completed.

"There were many issues with such a concept, child safety being paramount. Was the main spine of Walkodile® to be rigid or flexible, and if flexible, how flexible? How would the children be connected to it and how easy would it be to release them quickly if required? What happened if one child falls over, does it create the domino effect and they all fall over? These were just some of the issues that we faced" said Peter Ford and adds,

"But the single most difficult issue we faced was trying to ensure that the design remained within the claims of the patent that they had filed a lot earlier. This certainly restricted some of the things we could do with Elaine's idea. It would have been far better to have filed a patent application nearing the final prototype stage; this would have given us total freedom in design".

It would have also given John and Elaine a longer "golden window of covert activity".

John and Elaine had opted to get Walkodile® manufactured in the UK. As a pre-production prototype had now been completed by De Montfort University, tooling for the manufacturing and the national phase of the patent application had to be considered, but with no money left, where would they get the funding from to continue? Fortunately for John and Elaine, they had some very supportive relatives; Elaine's sister put in £50,000 and a brother in Canada put in a further £10,000. Supported by this funding, they were able to get the tooling completed and file their patent application in the national phase, covering several territories. John points out,

"We also felt our registered design was important in that the way our product looked needed to be protected almost as much as the function in the idea itself, so we made this a priority and also trade marked the name Walkodile®."

They incorporated the product into a newly formed limited company - Red Island Ltd. The first production Walkodile® was manufactured and John set about upping the profile of the product, entering it into various competitions; the awards were quite stunning!

The Awards

Walkodile®
Winner - British Invention of the Year 2007
Winner - UK Design of the Year 2007
Winner - Pan-European Shell Exploration Safety Exchange Competition 2005/06

Elaine Stephen
Winner - European Woman Inventor/Innovator of the Year 2007 (mother & child category)
European Special Recognition Award for Innovation 2007
Diamond Winner - UK Woman Inventor/Innovator of the Year 2007
Diamond Winner - UK Woman Product Developer of the Year 2007
Winner - UK Gold Medal for Outstanding Achievement in Innovation 2007

Red Island
Winner - Nursery Supplier/Innovator of the Year 2007
Winner - Grampian New Business Enterprise of the Year 2007

This was the springboard to achieving sales. Distributors were appointed in the UK and Europe to sell the product into education establishments although John admits that sounds far easier than it actually was,

"The interest has been phenomenal, yet when it comes to asking potential customers to part with money, there lies the problem; education authorities are strapped for funding so convincing them to buy Walkodile® instead of some other much needed element of education is difficult."

Yet despite this, John has persevered and the sales are coming in.

Asked if he would do it the same way all over again, John is quick to reply,

"No!" and qualifies his answer, "we filed a patent application far too early – this definitely put constraints on the final design; we should have filed in the later stages of prototype design which would have given us more freedom in the development of Walkodile®. The other point is a lot of money was spent early on at the working model and first, early prototype stages; if we had done far more research in finding an appropriate design institution for the job at that point, we would have saved absolutely thousands!"

John calculates that if marketing costs and his travelling costs are included with all the other costs for design, IP protection and tooling, the total expenditure to get Walkodile® to where it is today doesn't fall far short of £200,000.

Despite these costs, Walkodile® is starting to show signs of becoming a great success with sales increasing and John feels that now is probably the time to look at exclusive or non-exclusive licensing in the territories covered by the patent. He says,

"The IP protection is robust, the product is an excellent design as shown by all the awards and sales are increasing all the time, but it may be soon that we have taken it as far as we possibly can and we will soon be weighing up the options. Licensing Walkodile® presents an excellent opportunity I believe, not only for Elaine and I, but for whoever enters into a licence agreement as safety legislation is one of the driving forces and we have also proved that through the sales that we have achieved, there is a definite need for the product."

Looking at Elaine's idea and both her and John's efforts, they have turned what was a notion into future potential licensing deals in several parts of the world. Yes they have by their own admittance made mistakes and could have saved themselves maybe tens of thousands of pounds if they had approached the project in a different way. One thing they have done though, is turned what could possibly have been a "might like option" into a "must have need". It is all down to belief and not giving up even if the odds are stacked against you sometimes, but belief and effort are a pointless exercise in the development of your idea if you do not seek expert help – thankfully, John and Elaine did just that.

Chapter Fifteen

Conclusion

By now you will have given up any hope of turning your idea into a steady stream of income or be progressing steadily through the processes outlined in this book, ticking each box along the way or maybe your product is now in the latter stages of commercialisation. If it is the former, don't give up; another eureka moment will come along the way and then you can start all over again, using first time failure as a sure footing to progress that next idea.

It was never meant to be easy, developing your idea from a mere notion into a revenue stream, but the reward far outstrips the effort if you are successful. The saying that "patience is a virtue" never applied more than it does to idea development and mixed with "invitational" and your concerted efforts, there is no reason why your idea, if it jumps through all the hoops outlined in this book shouldn't be successful.

I wish you well in your endeavours and in short, never give up (even if your first idea fails), keep going and almost surely in time, the rewards will come. Oh, and finally, despite the odd bump or two along the way, enjoy the ride!

Rob Lucas

Chapter Sixteen

Training Courses

In the last few years, the government has invested a considerable sum in trying to turn the UK into a knowledge based economy. It is apparent that to do so, institutions and industry have to rethink their methodologies, particularly towards understanding and implementing innovation strategies and intellectual property management and development.

The author Rob Lucas undertakes training courses which address this "knowledge void". The courses are offered as either standard courses or tailor-made.

Standard Full Day Training

Standard Half Day Training

Both courses cover the contents of this book including practical examples, interactive tasks along with Q&A's, adding an extra and invaluable dimension to the understanding and managing of IP development and commercialisation.

Tailor Made Training Courses

As with any training course, the standard course may suffice, but there is often a need to tailor-make the content to suit the specific requirements and individual nuances presented by each type of industry or institution. The Standard Courses can be tailor-made for:

- **Universities and FE institutions**
- **Banking and the Finance Sector**
- **Business Advisors and Intermediaries**
- **Large Corporate Concerns**

If your organisation recognises training needs in innovation strategy and intellectual property management and development, or would just like to discuss the commercial, revenue generating opportunities presented by such training courses, please contact Rob Lucas at:

rob.lucas@jembro.org

Further information on Jembro training courses can be found at:

www.jembro.org/training.aspx

Chapter Seventeen

Institution Details

The contact details of the institutions in this book are shown below should you require any further information.

The British Library
96 Euston Road
London
NW1 2DB
United Kingdom

Switchboard	+44 (0)870 444 1500
Website	www.bl.uk
Business Centre	www.bl.uk/bipc
Research Office	research@bl.uk

De Montfort University
Faculty of Art and Design
De Montfort University
The Fletcher Building
Mill Lane
Leicester
LE1 9EB
United Kingdom

Telephone	+44 (0)116 250 6238
Website	www.dmudesign.co.uk

UK Intellectual Property Office
Concept House
Cardiff Road
Newport
South Wales
NP10 8QQ
United Kingdom

Telephone	0845 9 500 505
Website	www.ipo.gov.uk

Chartered Institute of Patent Attorneys
95 Chancery Lane
London
WC2A 1DT
United Kingdom
Telephone
Website

+44 (0)20 7405 9450
www.cipa.org.uk